Crab and lobster
fishing

Dedication

Dedicated to all those who wrest a living from among the rocks and stones of the low water mark; the Crab and Lobster Fishermen.

Crab and lobster fishing

Alan Spence

Fishing News Books Ltd
Farnham · Surrey · England

© Alan Spence 1989

British Library Cataloguing in Publication Data

Spence, Alan
  Crab and lobster fishing.
  1. Great Britain. Coastal waters. Crabs.
  Lobsters. Crabs & lobsters. Fishing
  I. Title
  639'.54

  ISBN 0 85238 154 9

Published by
Fishing News Books Ltd
1 Long Garden Walk
Farnham, Surrey, England

Typeset by
Mathematical Composition Setters Ltd
Salisbury, Wiltshire

Printed in Great Britain by
Henry Ling Ltd
The Dorset Press, Dorchester

Contents

Illustrations

Preface

Catching shellfish by baited traps has been practised for very many years in the coastal regions of many countries of the world. Traps were generally constructed from local materials, and techniques were developed by trial and error. The development of synthetic materials has introduced changes in the design and production of traps although many fishermen still prefer to construct their own. The introduction of powered hauling gear and electronic aids is also now exploited in this form of fishing.

Having a life-long interest in fishing, particularly in the sea around the coast of Southeast Scotland, and subsequently gaining experience with one of the best lobster fishermen in the area, my full-time involvement in the catching of crab and lobsters became inevitable. The life can be very hard and difficulties and disappointments are part of it. However there are the good times when no other independent form of livelihood can be considered.

Although this book is based on my experience as a professional crab and lobster fisherman around the Scottish/English borders it may have application and benefit for others involved or wishing to participate in this form of fishing. One point I would make is that some experimentation is always worthwhile with gear, as with bait and its presentation.

Alan Spence
March 1989

Acknowledgements

My thanks go to all the scientists who replied to my letters of enquiry, the granting of permission to cite their work and the Fishery Officers around the coast who filled in and returned the questionnaire on the main seasons for crab and lobster landings. Also the late Arthur Collins of Burnmouth and his wife Jenny for the many happy hours spent in discussion of 'the fishing' past and present. Not least Mrs Rita Davidson who transformed a bundle of dog eared sheets of badly typed notes into a presentable manuscript.

Introduction

Here in Britain fishermen are well served by a number of official bodies connected with the fishing industry, part of whose duties is to study the biology, catching and marketing of lobsters and other shellfish. Reports of these studies and other experiments are published by Department of Agriculture and Fisheries for Scotland (DAFS), Ministry of Agriculture, Fisheries and Food (MAFF) and Sea Fish Industry Authority (SFIA). These reports are of a highly accurate although often involved technical or scientific nature, their findings backed up by appropriate graphs and tables. Graphs and tables are something which, as a fisherman, I seldom read in any publication and depend very much upon the written text for information.

In this book every effort has been made to keep graphs and tables to a minimum, only using them where really essential. Hopefully by marrying the quoted scientific works with personal experience and the experience of other fishermen, the finished book will be readable as well as informative.

Where description has been difficult in the written word diagrams or photographs have been provided rather than a long drawn out text.

Where reference is made to peak fishing times such as after the emergence of lobsters from moulting this is specifically for the east coast of Scotland and North East England. In the crab fishery there is a wide difference in peak times even between the cock and hen, to attempt to list them all would have been futile. For those who wish to explore further the scientific research which has been carried out on crabs, lobsters and their fishery, a list of publications appears at the end of this book.

1 Basic biology of crabs and lobsters, their habitat and fishery

Away in the distant past, even before man had discovered how to make the most primitive of vessels he found that shellfish could make a welcome change to his diet. Among the middens of ancient settlements near the sea there are often found the remains of molluscs and other bivalves. When collecting these from the intertidal zones at low water there is no reason to doubt that early man also captured both the edible crab (*Cancer pagurus*), and the common lobster (*Homarus gammarus*), during his pryings and pokings among the tide edge rocks.

While both these species have a wide distribution from the Arctic Circle to the Mediterranean, by far the greatest numbers are found in what could be best described as Middle European waters.

Lobsters are to be found in Norwegian, French and Irish waters but by far the greatest concentrations are in the coastal waters of the UK.

The edible crab has a similar geographic distribution as the lobster but in this case the catch is more divided with Norway, Britain and France being the main participants.

In the UK the main effort for crab fishing has historically been concentrated very much on the east coast from Cromer north to Wick. Other fisheries exist in the English Channel extending westwards from Selsey to the grounds southwest of Cornwall and around the Welsh coast.

Previously unexploited grounds in the northwest of Scotland are now being fished for crabs, where previously they were regarded as something of a pest species when they were caught in traps aimed for lobsters.

Fig 1 The type of foreshore most favoured by lobsters

'The lobster is completely enclosed in an exoskeleton (outer shell) of calcified chitin. The body is composed of combined head and thorax enclosed in a carapace and an abdomen of six articulated segments. The most easily recognised characteristics are the large claws (chelae), used for tearing and crushing food. One of these claws has serrated edges, which overlap scissor fashion and are used for cutting and tearing. The other claw is much more heavily built and the two edges meet directly together to form a very efficient crusher. The first two pairs of walking legs also have small pincers which are used to pick up food and pass it

14

to the mouth parts. The mouth parts cut and tear the pieces of food into small particles which pass into the stomach where further maceration takes place in the gastric mill by the stomach teeth (ossicles).

Fig 2 Lobsters face any potential danger with their best defence – powerful claws

'The lobster also has a long pair of antennae which are sensitive to vibrations and shorter antennules which detect chemicals. Compound eyes lie either side of the rostrun. Lobsters usually walk but can escape rapidly backwards by sudden flexing of the abdomen. The sex of a lobster can be determined by the different shape of the sexual appendages, sited underneath the abdomen. In addition the male lobster has larger claws and a narrower abdomen than the female.'[1]

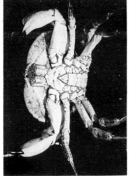

Fig 3   Male crab

Fig 4   Female crab

As in the case of the lobster the sex of the male crab is at once obvious by the greater size of the claws. Unlike the lobster both claws of the edible crab are of the same construction, the crushing type. Also, the external sexual distinguishing features are more apparent than in the case of the lobster. (*Figs 3* and *4*) A further guide is in the shape of the shell, that of the female being much more dome shaped than the male's.

15

Habitat   To a large extent adult crabs and lobsters share a
similar habitat for the greater part of the year. In the
case of the latter they are almost exclusively to be
found upon a sea-bed which is rocky, with reefs,
boulders and large stones or wreckage forming the
main shelter for the fish. Many edible crabs are also to
be found on ground such as these but can also be
taken on sand, gravel and even clay or mud bottoms
depending upon the season. Sea-bed forests of kelp and
weed can at certain times be much favoured by crabs
and lobsters even although sea-bed motivation must be
difficult in the thicker growths.

   While the rough broken ground is the main habitat
of the mature lobster, observations by the Ministry of
Agriculture Fisheries and Food (Howard and Bennett
1979[3]) indicate that juveniles will construct tunnels in
mud in the same manner as the Norway lobster
(*Nephrops norvegicus*).

   It is the nature of the sea-bed, and the size and
number of refuges available, which is the governing
factor of the composition of the lobster stock on any
given ground. Where there are large boulders and
consequently large refuges there is every likelihood that
these will be occupied by large adult lobsters.

Fig 5 Ground suitable for
        crabs and lobsters

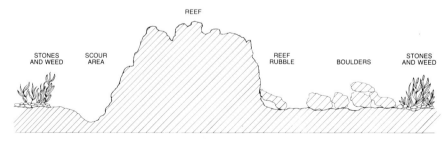

REEF

STONES
AND WEED   SCOUR
            AREA   REEF
                             RUBBLE   BOULDERS   STONES
                                                         AND WEED

Fig 6 Examples of mixed
hard and soft ground as
shown on graph-recording
echo sounder

(A) AT X the small patch
of weed shows up
more clearly on the
second echo than the
first. Y indicates kelp
and weed over hard
ground. Z, the loss of
the second echo
indicates that there is
soft ground here

A

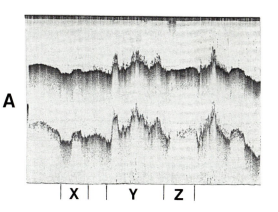

60 ft

| X | | Y | Z |

(B) X and Y again show
on the second echo
that this is sand or
mud

B

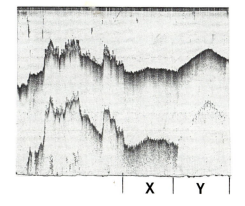

60 ft

| X | | Y |

(C) Although X appears
flat it is according to
the second echo in
fact hard, or at least
stony, when compared
to Y

C

60 ft

| X | Y |

17

'*The selective influence of lobster survival in an area of (a) crevice availability (b) volume of lee from water currents*

(a)   Cobb (1970) and Dybern (1973) have shown that for *H. americanus* and *H. gammarus* respectively that a lobster occupies a hole proportional to its size. It can be readily appreciated that in a given area of sea-bed many small rocks will offer more small crevices suitable for small lobsters than a few large rocks. Conversely a few large rocks will offer more large crevices suitable for large lobsters than many small rocks. Thus the relative availability of shelters could be an important control on the numbers of lobsters of any size group that could survive in an area.

(b)   Observations in a flume (Howard and Nunny, 1983[4]) have demonstrated that the locomotory activity of lobsters is limited to currents of less than 25 cm (9·8 in) sec^{-1}. Lobsters occur commonly in areas experiencing surface tidal flows of 125 cm (49$\frac{1}{4}$ in) ft/sec^{-1}, which implies a current of 50 cm (19$\frac{1}{2}$ in) sec^{-1} at 5 cm (2 in) above the bottom. It appears likely therefore that lobsters need to exploit the areas of lee created by outcrops on the sea-bed. As the volume of the lee produced by an outcrop is proportional to the cross section area it presents to the current, it seems likely that the size of lobster able to shelter in the lee will also be proportional to the size of the outcrop. Flume experiments (Howard and Nunny 1983[4]) suggest that this is so. For example juvenile lobsters (12–25 mm ($\frac{1}{2}$–1 in) CL) were able to shelter behind a block 22 × 21 × 13 mm ($\frac{7}{8}$ × $\frac{7}{8}$ × $\frac{1}{2}$ in), in a current of 55 cm (21$\frac{1}{2}$ in) sec^{-1}, whereas "adults" (70–90 mm (2$\frac{3}{4}$–3$\frac{1}{2}$ in) CL) would not.

'It would seem reasonable to suppose that as an individual lobster grows and becomes dissatisfied with its niche that it moves during slack water to a more suitable site. The size distribution of out-crops (and hence niches) on the sea-bed has a pattern created by transport, erosion, etc. Therefore, when the whole population is making this niche selection, one would expect that the lobster size distribution should reflect, to some extent, the sea-bed topography.'[2]

Ecdysis Both crabs and lobsters have their bodies enclosed in a hard exoskeleton or outer shell. Rigid and unwielding there is no provision for growth within this shell and size increase can take place only by moulting or casting the old shell.

'This process is known as moulting or ecdysis. Lobsters tend to moult at certain times of the year, but the moulting period may take place at different times on different parts of the coast. Prior to each moult a new shell develops beneath the old one before it is cast. When moulting is about to take place the lobster usually retreats into hiding because after moulting the lobster will be soft shelled and defenceless and easy prey for fish such as cod, dogfish and conger eel. At the start of moulting the membrane between the carapace and abdomen splits; the soft-shelled lobster then "jack-knifes" out of the old shell by a series of contractions. The first part of the soft lobster to emerge from the old shell is the area at the back of the carapace and the top end of the abdomen, followed by the legs, claws and finally the abdomen. Not only the complete shell but also the lining of the gut, stomach and gills are cast. Because the new limbs are soft and pliable they can be pulled through the joints of the old limbs.

'After emerging from its old shell the lobster is very soft wrinkled and pliable; it then swells to the new size mainly by the intake of water, the new shell gradually hardens by a build up of calcium salts. The shell will become rigid within a few days but will not be completely hardened for about three

19

or four weeks. No further increase in size will occur until the next moult. Before and after the moult there is some suggestion that a lobster will change its diet to one rich in calcium. Lobsters which moult when kept in captivity usually eat their old shell, even when kept supplied with ample food. This is probably in an effort to obtain calcium required to harden up the new shell.

'Lobsters are tagged with a special tag, inserted in the pliable membrane between the carapace and the abdomen, which is not lost when the lobster moults. It has been found that on average lobsters increase their carapace length by 10% and weight by 50% at each moult. It is not possible to state the age of a lobster with accuracy as no part of the body gives a reliable guide, all hard parts which might show ageing being lost at each moult. It is estimated that a lobster takes between four and six years to reach the minimum legal landing size of 80 mm [3·15 in] carapace length which is equivalent to a total length of 229 mm [9 in] and an approximate weight of 340 g [12 oz]. (The minimum legal landing size has since been increased to 85 mm [3·34 in].)'[1]

In a similar manner the crab must follow the same ecdysis process as the lobster but in this case it may be many months rather than weeks before it is in marketable condition.

'The crab is encased in a hard, rigid shell of fixed shape which must be cast off at intervals and replaced by a larger one so that the crab may have room to grow; this process is known as moulting or ecdysis. In Britain the main moulting period is from July to October; the females start moulting in July, followed by the males a month or so later. At this time the crab seeks shelter among the rocks. The shell cracks along a precise line dividing the upper and lower halves, and the soft crab inside slowly backs out through the gap. It then absorbs water and swells, increasing in size across the back by as much as 20 to 30% in one moult. Males appear to grow slightly more during a moult than females, but moult less frequently after they reach a size of $4\frac{1}{2}$ inches. On average a $3\frac{1}{2}$ inch male crab on the east coast will reach $4\frac{1}{2}$ inches in one moult and $5\frac{3}{4}$ inches on the next.

'After the moult has been completed the shell slowly hardens and no further increase in size will occur until the next moult, although the shell will not be completely hardened for two or three months. Moulting takes place at frequent intervals during the crab's early life, but after it has reached a size of 5 inches moulting takes place about once every two years. It is not possible to state the age of a crab with accuracy, since no part of the body gives a reliable guide, but on average a $4\frac{1}{2}$ inch crab is about four to five years old. Most female crabs are mature by the time they measure 5 inches across the back, whilst males reach maturity at a slightly smaller size.'[5]

The emergence of lobsters after ecdysis

In some areas the appearance of lobsters after ecdysis is an amazing event. Over a period of weeks lobsters have become increasingly scarce on the grounds as more and more of them seek secure niches for the ecdysis process. One day baited traps may be taking only the occasional clean moulted lobster, these are easily recognised by the absence of marine growths such as barnacles on their shells, a slight 'oily' feel and movement of the carapace when picked up. Suddenly there is an explosion of clean moulted lobsters on the ground with the catch rate doubling every day until a peak is reached.

In other areas the emergence is less dramatic being a more gradual spread out process. Throughout Britain the key month for lobsters completing the process would appear to be August, although in some districts it can occur as early as May. (See map and key which gives these times and peak periods of fishing for both lobsters and crabs [Appendix 1].)

Lobsters occasionally moult within creels, apparently entering them in the search for a secure cranny when at their most vulnerable. In this instance the old shell can be found in the creel beside the newly moulted lobster. Or a lobster may be found in the creel, apparently newly moulted and jelly like but without its old shell. In this instance the lobster may either have eaten the shell or it has been washed from the creel. In either instance the fish is valueless, unlikely to survive unless returned immediately to the sea.

While peak landings of lobsters occur mainly in the post ecdysis period on some parts of the coast, a spring fishing exists where several weeks of good fishing may be had from shallow water – presumably as the lobsters move inshore to moult.

In general the first three months of the year are the least productive for the crab and lobster fisherman. Low water temperatures causing the shellfish to become lethargic, and the threat of bad weather preventing the most productive grounds being worked cause some fishermen (particularly those who operate with small vessels) to take their gear ashore for these months.

At the start of the lobster season after ecdysis the largest catches are often taken by small shallow draughted boats working close inshore among the drying boulders and rock edges. This would seem to indicate that to escape any potential predator the lobster likes to cast his shell around or even above the low water mark. As it is possible, where a suitable foreshore occurs, to collect lobsters from holes and crevices well above low water mark this could be true.

This habit of the lobster in seeking refuge for ecdysis so close inshore is the main reason why when weather conditions permit, fishermen will shoot their gear as close inshore as possible. It is at this time that the small boat with a one- or two-man crew reap their harvest, picking up the lobsters as they emerge. Frequently their catches will exceed those of larger, more sophisticated vessels which due to their draught may be unable to penetrate any farther inshore than depths of three or four fathoms.

Reproduction process

As the process of ecdysis follows similar lines with both crabs and lobsters the reproduction process is also shared by both species. In both, mating can only take place between adult females immediately after they have shed their old shells, and males still in the hard state.

'Mating occurs in inshore waters during the summer, immediately after the female crab has moulted and while it is in the soft-shelled condition. Prior to the moult and for a period of up to a fortnight after, the female is attended by a hard shelled male. Immediately the female has cast her shell mating takes place and the sperm is introduced into the female's sperm sacs. One supply of sperm may fertilize two or more batches of eggs in subsequent years, and the majority of females which mate in July or August will spawn, carry eggs, in November and December of the same year, but in some cases spawning is delayed until the next winter. Crabs usually select a soft sea-bed for spawning, often in deep water, and the eggs remain attached to the swimmeret on the abdomen of the parent for about seven months. A crab with eggs is called a berried crab and the number of eggs carried can vary from half a million on a 5 inch crab to three million on a 7 inch one.

'In the spring and summer following spawning the berried females move inshore, where the eggs hatch: the young crabs at first have a shrimp-like appearance and form part of the free floating plankton. This period is believed to last about a month, and during this time the larval crabs will probably drift to new grounds away from the hatching area. It finally settles on the sea-bed and assumes the adult form when it is about $\frac{1}{8}$ inch in size.'

'Tagging experiments have shown that on the east coast of Britain mature female crabs can move considerable distances, mainly in a northerly direction. Crabs released off Whitby, Yorkshire, have been recaptured along the Scottish coast, having moved between 180 and 200 miles in twelve to eighteen months. These migrations are associated with the offshore movement of females for spawning. Male crabs rarely move far from their release point.'[5]

'Mating takes place between a soft-shelled female and a hard-shelled male. Although there are records of mating when both sexes are in a hard shelled condition it is not certain whether fertilization takes place during these activities. The sperm is retained in the receptacle of the female until the eggs are laid after about four months, but the interval can be longer. The eggs are fertilised externally as they pass out from the oviduct openings on the basal segments of the third pair of walking legs. They then become attached to the hairs of the abdominal appendages or pleopods. When the eggs are laid the female lies on her back and curls her tail forward and extrudes the eggs on to the pleopods. They remain attached to the pleopods for approximately ten months, being constantly aerated and cleaned by the female. The majority of females carrying eggs (*i.e.* berried) have a carapace length above 80 mm (3·15 inches). A berried female will carry between 5,000 and 100,000 eggs depending on her size.

'The eggs usually hatch about June or July and the young lobsters are released into the plankton to fend for themselves. The lobster larvae then moult

Fig 7 The female lobster carries her mass of berries safely below the abdomen

24

three times, changing their form with each moult. After the third moult (stage IV) the larvae spend most of their time on the sea-bed and after the fourth moult (stage V) assume the shape and characteristics of the adult lobster. The time taken for the lobster larvae to reach this fifth stage is dependent upon water temperature and the availability of food. Neither larvae or very small lobsters are easily found in natural conditions and certainly not in sufficient numbers to account for the numbers of commercial lobsters caught. Recently catches of small numbers of larvae have been found in plankton by use of a neuston net. These larvae were mainly stage I, but stages II, III and IV were also found. Laboratory experiments with juvenile lobsters suggest that they may differ in their habitat preference to adult lobsters, and will for example form quite intricate burrows in mud, where they spend most of their time. If juvenile lobsters have a different habitat and therefore possibly a different food preference, this fact could account for the absence of very small lobsters below 4 inches (102 mm) in total length in commercial catches.'[1]

Problems of rearing Bearing in mind the already mentioned need for lobsters to have both cover and shelter from sea-bed currents any action to increase the stock of lobsters on the ground has several problems to surmount. There would be little point in hatching lobsters to the IV larva stage and simply casting them on the water.

Problems also exist in the rearing of lobsters as throughout their lives they have an unfortunate tendency to eat those of their own kind. This means that any rearing programme which goes beyond the egg stage must have provision for individual compartments for each lobster in the rearing tanks.

At the moment the Ministry of Agriculture Fisheries and Food are conducting field trials on lobster 'seeding' with reared lobsters of around 3 cm (1·18 in) overall length. These trials are taking place in Bridlington Bay in Yorkshire with the small lobsters being released on the sea-bed where the topography is suitable. Releasing is done by divers adjacent to suitable cover, which the juvenile lobsters soon adopt.

Possibilities include the implanting of magnetic tags in the released lobsters to determine the number of these lobsters taken commercially.

Autotomy

Autotomy, the ability to regrow lost limbs, is a phenomenon shared by both crabs and lobsters. Both can shed limbs voluntarily when seized by a predator or caught in an obstruction.

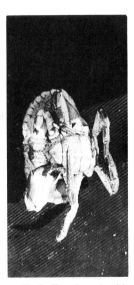

Fig 8 Female crab with regenerating claw

'Autotomy is the process whereby an injured limb can be shed allowing a new one to be regenerated to replace the old. In some species, such as the lobster, the process of autotomy is used to enable escape from a predator. This is particularly the case if the lobster is seized by the claws; it can shed them and escape by vigorous movements of the tail. When a limb is autotomised it is separated at a fracture plane at the base of the limb; this fracture limb is covered by a thin membrane perforated by a small hole. The blood flows out through this opening but soon coagulates and seals the membrane. As the new limb is regenerated it grows out of the old stump and grows to a size slightly smaller than the limb which is lost. At this premoult stage the limb is soft and enclosed in a membrane. It will not harden up until the lobster moults. Occasionally a limb, usually a claw, will become stuck when the lobster moults and the lobster will release itself by shedding the new limb.'[1]

Predators

On the sea-bed, shellfish such as crabs and lobsters have many enemies and hungry mouths ready to feed upon them. Lobsters will kill a soft crab within the confines of a creel or pot but whether such a thing

26

occurs in the wild is a matter of conjecture. Most likely individuals of each species when of similar size will treat each other with respect, but they will equally take every opportunity to feed on smaller individuals, be they crabs or lobsters.

Jungle conditions exist on the sea-bed; it can be the survival of the fleetest, fittest and best hidden. Octopuses, which have the same range of latitude as lobsters, seem partial to the latter's flesh. As they can kill and extract the flesh from a lobster through the meshes of the net covering of a creel, it is equally possible that they can reach into a crevice to do the same thing. Conger eels and seals are main predators, but in some stage of the lobster's life from egg to adulthood they are vulnerable to predation by a wide variety of sea fish, mammals and other crustacians.

Codling of five pounds have been found to contain lobsters of the, at the time, legal landing length of 12 inches overall length. At the other end of the scale codling caught on lines over rough ground during the summer months contain in their stomachs what appear to be large numbers of 'ripe' lobster berries. Whether this is what they are or not has not been ascertained by scientific evidence, but in Northumberland it is the belief of many fishermen that codling actually remove the eggs from the berried females.

Fig 9 Octopus: a lobster predator

Diseases

Few creatures are free from any disease and lobsters are no exception in this case. Several affect lobsters, some of which merely affect the shell while others are fatal to the lobster.

'Although lobsters probably suffer from a number of diseases, many of them parasitic, there are only a few which affect the economics of the fishery. Some lobsters have a blackening of the shell caused by bacteria which break down the chitin of the shell; the flesh is not normally affected except in heavily infected cases. The main disadvantage of shell

27

disease is one of appearance and the lobsters affected usually fetch a low price. Most cases of shell disease, unless very severe, clear up when the lobster moults. When lobsters are kept in large quantities in crowded conditions care must be taken to ensure that no disease affects them. The best method is to keep the tanks and water clean, the water cool and well aerated and not to overstock the tanks. The disease most likely to show itself in bad holding conditions and cause heavy losses is gaffkaemia, a disease of the lobster blood caused by a bacterium, *Aerococcus viridens*. A recent survey has shown that its incidence in England and Wales is low at the moment. Large losses of lobsters in holding tanks are usually due to oxygen depeletion caused by rising temperature, poor water circulation or overstocking.

'A copepod parasite, *Nicothoe astaci*, is sometimes found on the gills of lobsters, but this has little effect on the lobster and none on its value.'[1]

Effect of cold weather   Other factors can cause the death of lobsters in the wild. The early months of 1986 were marked by a prolonged spell of cold weather in the southeast of Scotland and Northumberland. Snow fell on higher ground inland from January to April resulting in large discharges of cold fresh water into the sea from major rivers in the area.

During late May and June, when sports divers were operating around the Farne Islands in Northumberland, they found 'dozens' of lobsters dead on the sea-bed. Further investigation found that many were also dead in their holes and crevices. Following a spell of easterly gales lobsters were also found washed up dead on the shore in numbers never before experienced.

Dead lobsters were sent to the Department of Agriculture, Fisheries and Food for post-mortem examination to ascertain the cause of death. While no

28

definite cause of the demise of such a large number of lobsters could be proven, the specimens examined proved to be free from disease. As no stomach contents were found in the lobsters examined the conclusion was reached that it had been the prolonged period of weather conditions which had been the cause of this disaster. When scientists gauged the temperature of the sea at the Farnes in June it was found to be that which would be normal in December.

Another circumstance of weather which can kill lobsters is a long period of heavy seas when they are close inshore for the ecdysis cycle. Such weather in autumn can result in lobsters being killed by breaking seas and getting washed ashore among uprooted kelp.

However, many animals seem to have an inbuilt sixth sense which warns them of impending change in weather conditions. In this lobsters are no exception. A catch of lobsters far above the seasonal average or exceeding the previous day's catch by a large amount usually means the onset of sea winds within two days. If this catch is followed by one which is extremely poor, a blow is all but guaranteed for the following day. Whether in these circumstances shellfish are feeding before they take refuge, or they enter the creel in search of shelter is uncertain. When catch rates such as the above occur it is advisable to move gear to the safety of deep water.

Choice of location by crab and lobster

As has been already quoted, lobsters and to a lesser extent crabs, depend very much upon the shelter and lee provided by a broken sea-bed. This seems to bear out the experience of many fishermen in that the knowledge of the location of suitable reefs and pinnacles is not always sufficient for successful operation. On reefs which run across the direction of the tidal flow, traps shot on one side may not have been catching, while those on the other have been fishing well – a situation which very much depends upon which side of the reef happened to be the lee or sheltered side during the hours of darkness when lobster activity is greatest. Slack water during the dark period might find that creels shot each side have fished equally well, with the prolific side of the reef changing with the altering cycle of tide and darkness.

If the movement of lobsters is hampered by strong sea-bed currents, the crab with its low-slung shape is less so. In fact it is apparent to most fishermen that the crab is stimulated to feed and thus enter creels by strong tidal flows. On many grounds crab catches are always better, all other factors being equal, during spring tides rather than neaps. Why this should be is difficult to ascertain. With a voracious appetite compared to that of the lobster, it may be the case that stronger tides carry the bait scent, and draw crabs from a greater distance.

'In the wild the diet of crabs and lobster consists of worms, molluscs, fish and starfish. Despite both crabs and lobsters being taken in traps, sometimes in the case of the latter by bait which is far from fresh, in the wild their diet consists mainly of living prey. For both species this will be crustaceans, marine worms, molluscs and fish.'[5]

Appetite stimulated after ecdysis   Feeding activity for both crabs and lobsters is at its greatest when they emerge from ecdysis then wait for the new shell to harden. Not only has the weeks without food made them ravenously hungry, but within the new shell there is vacant body space to fill with flesh which can be done only by eating. Another factor which must be taken into consideration is that shellfish activity and therefore feeding is at its height in the warm waters of late summer and autumn. Crabs and lobsters do not feed very well when water temperature falls below 5°C. Taking the above two factors of hunger and suitable water temperatures into account it is not surprising that in inshore waters at least, the majority of the lobster catch is made from August to November.

Poor condition of crabs after ecdysis Considering the time span required for crabs to regain condition after ecdysis, very few of these autumn crabs which have moulted will be in marketable condition.

These poorly conditioned crabs are protected in some sea fishery areas by a ban on landing or offering for sale 'white toed crabs' this state being an indication that the crab had not regained condition following ecdysis. While this is a general pointer of the crab's condition, there are many crabs whose claws 'toes' have turned the normal black colour, but which yield little or no meat.

Where there has been little pressure on the edible crab stock through fishing effort the effects of over-population can lead to a high proportion of diseased shellfish in the catch. Poor shell condition is the prime indicator of this condition and in extreme conditions anything up to two thirds of the catch must be discarded or destroyed.

Few lobsters are taken in traps which are shot and hauled during the same period of daylight. It can safely be assumed that lobsters are creatures of nocturnal habit, only moving from their holes and feeding during the hours of darkness. Depending upon the senses of smell and touch more than that of sight, there seems little handicap to the lobster when feeding during night time. The large mass of bait within a trap would be easily located by a lobster compared to its usual diet of small shellfish and worms.

The role of the creel in attracting lobsters 'The assessments we make of the lobster stocks around the coast of Scotland are based upon samples taken from the commercial creel fishery. It matters a great deal to us to know how representative these samples are, of the stocks of lobsters in the sea. We know from the work of a former colleague,

31

Dr H J Thomas, that the creel is a rather selective fishing gear, small lobsters may enter and leave the creel more readily than their larger brethren and the largest lobsters may be excluded altogether if a normal "hard" eye is fitted.

'The dimensions of the creel may not be the only selective forces at work. On the better stocked grounds, some selection might take place even before the lobsters enter the creels. To investigate this properly it is necessary to follow the whole sequence from a sheltering lobster's first awareness of a baited creel to its successful capture. We began by watching the behaviour of lobsters in large aquaria. We soon confirmed that lobsters are most active when light intensity at the sea-bed is comparatively low. In full daylight lobsters spend most of their time concealed in whatever shelter is available. In the normal concealment posture, the outer entrance to the shelter is guarded by the large claws with the tail pointing inwards. There is a group of light-sensitive nerve cells in the end of the tail. The function of these cells is presumably to warn the lobster if the vulnerable tail is projecting out of the other end of the shelter. In our aquarian experiments, we used old drain pipes to provide the necessary "accommodation". We found that, just as a hermit crab selects a tight fitting shell, so the lobster tends to choose a shelter which fits it reasonably closely. It seems that in concealing itself, the lobster "likes" to have as much of its body and tail in contact with a solid surface. Although quite hard, the cuticle of the lobster is highly sensitive to touch. This is partly because of its many mechanically-sensitive hairs but probably the most important of the true touch receptors are special disc-like organs which respond to minute strains in the cuticle much like the strain gauges in the skin of a modern aircraft.

'When a baited creel was first put into the

Fig 10 Much of the lobster's awareness of its environment depends on signals received by the sensitive antennae

aquarium, the lobster made no response apart from a deeper withdrawal into its shelter in response to what seemed to be a "frightening" disturbance. After some minutes, depending upon the circulation of water in the aquarium, the small antennae of the lobster increased their rate of beating and the lobster left its shelter. Experiments in which the nerves leaving the antennae were connected to powerful amplifiers showed that the outer branches of the small antennae are liberally provided with hairs which respond to minute quantities of dissolved chemicals. Each hair contains two to three hundred nerve cells. Stimulation of the hairs by dissolved chemicals washing out of the bait is the trigger which initiates the walking response. The overall path followed by the lobster is up the current but the detailed track may be quite tortuous. As the exploratory walk proceeds and the lobster encounters higher concentrations of dissolved chemicals so other sensory hairs are stimulated on the outer mouth parts and walking legs.

'Finally, the lobster encounters the creel. To the lobster the apparent sources of the attractive scent is not always the bait inside the creel but the string with which the creel is closed. Very often this has been tied with bait contaminated hands and the newly-arrived lobster will spend many minutes picking away at it with the small claws on the ends of the first pair of walking legs. Contaminating the closing string with bait is clearly something to be avoided when setting creels. Eventually the lobster loses interest and walks around and over the creel, often taking a long time to find an entrance. It may eventually be successful and, if so, usually enters with the claws leading. I heard of one veteran lobster fisherman who used to advertise the entrances by wiping them with bait-contaminated hands which, in the light of what we have seen, appears to be a sound idea.'[6]

33

Varying locations of catches As has been mentioned earlier in this chapter the catch of lobsters immediately after ecdysis is usually best in shallow water close inshore. As autumn progresses the bulk of the catch will be taken farther and farther offshore, and in correspondingly deeper water. Whether this is as a result of lobsters which have moulted inshore moving off into deeper water, or of an emergence of lobsters which have moulted in deeper water is uncertain. As autumn merges to winter fishermen will be shooting their creels in deeper and deeper water, yet most years when weather conditions permit, catches can still be made among the drying rocks at low water. In mid-winter it can be a case of 20 fathoms or almost zero, lobster from depths in between being few and far between.

The bulk of the marketable crab catch is taken in the opposite circumstances when the crab is in its pre-ecdysis state. During the winter as water temperatures drop the crab's activities have become increasingly lethargic. For example in the late cold spring of 1986 the few boats in north Northumberland and Berwickshire which continued to fish during this period were hauling 200 or even 300 creels once every two days and failing to land one stone of crabs.

Traditionally fishermen have talked of the crabs being 'sanded up', *ie* buried in sand during periods of low sea water temperature. While this may not be a scientific fact what is certain is that when water temperatures are low few crabs are inclined to feed and enter baited traps.

As the daylight period lengthens in spring and the water temperature rises the female crabs begin their inshore migration (see page 23) to allow the hatching of the eggs they have spawned seven months previously. It is at this time of year that the best catches do not necessarily come from the rocky hard ground usually fished. Many of the largest catches of spring and summer crabs can be taken over sand or mud in depths of water between two and five fathoms. While it is not always the case, some of the largest concentrations are to be found along the edges of hard ground, in particular where any scouring by tide action has created a pit or hollow in the sand.

The main inshore migration of male crabs is usually two or four weeks after this, presumably the males are awaiting the ecdysis of the females, allowing mating to take place.

The working of baited traps

Like many other methods of fishing the working of baited traps for shellfish is an early morning job. There is a twofold purpose to this, one being that while the catch will be content within the creel during the hours of darkness or when the sun is low in the sky, daylight and a high sun overhead may make the catch restless and try to leave the creel or pot. This may not have much effect in deep water, or in the winter months, but in summer it is a factor worth taking into consideration. The other point is that hauling baited traps is a physically active job, it is also a wet one with fishermen encased in waterproof clothing. At noon in the height of a summer day large amounts of moisture are generated within the oilskins so it is much more comfortable to work in the cooler conditions of early morning.

Around the coast many small harbours and communities depend entirely upon the revenue generated from crab and lobster fishing. Without the income from the trap fishery small harbours could not be maintained and many coastal villages would go the way of other rural communities and degenerate into further enclaves of holiday homes and 'dormitory' towns.

Fig 11 Lobster fishing is an early morning job; sunrise off the Berwickshire coast

The effect of the crab and lobster on the economy

Statistics are available from official sources which give the importance of crab and lobster fishing both locally and nationally as a foreign currency earner. The Sea Fish Industry Authority, DAFS and MAFF and other organisations can supply charts, tables and diagrams

35

Fig 12 Small harbours like Burnmouth depend upon crab and lobster fishing vessels for survival

showing landings for crabs and lobsters in weight, cash value terms and export earnings for those interested.

Statistics are usually outdated by the time they are published, although with the modern trend to record this data on computer disc the delay will be reduced considerably and there will be up to date, rather than recent statistics.

That is however beside the point, for example in 1984, the most recent statistic, about a quarter of the UK crab catch was exported, while three-quarters of British lobsters were sent to destinations outwith the UK.

Static gear fishing compared with mobile

Some big boat men look with scorn on the work of the trap fisherman, saying it is an 'old man's job'. Certainly the hours spent at sea can be less, but the constant need to make and maintain gear ensure that the hours are as long as any in the fishing industry. Working any form of static gear is very different from that of mobile fishing such as seine netting or trawling.

His gear safely stowed on board, the trawl or seine net man can lie snug abed as the wind gusts around the walls of his home. With several hundred vulnerable

36

creels in the sea, the creel fisherman awakes and wonders at the direction of every blast. A changing wind overnight blowing from the sea means that if possible the shellfisherman must put to sea in an attempt to move his gear offshore.

Expenses and profits shared by crew

Traditionally, inshore boats are crewed on a share basis, this can be with several of the crew owning shares in the vessel or working only for a share of the profits from the catch. In this the shellfish section is no exception, with the majority of crews working on the basis of no catch no wage.

In most instances the boat will provide the rope anchors and dahns, although there may be a weekly deduction to cover this expense. When working on a share system most fishermen are expected to provide and maintain a proportion of the traps to be fished by the boat.

Where boats are company owned rather than by the crews the gear is likely to be supplied with the proviso that the crew must repair and maintain them.

The changing of traditional fishing methods

Over the past three decades many of the traditional methods of inshore fishing have fallen into decline in some areas. One of the few to survive universally around the coast is the catching of shellfish in baited traps. In Scotland the art of 'small' lining was largely replaced after World War II by the use of the seine net over soft ground. This in itself, has now been displaced to some extent by the use of single boat and pair trawling.

Another one time important fishery which has also been almost abandoned is the drift net fishery for herring. Today the drifters have been replaced by purse seiners and mid-water trawlers whose catch is measured by the tonne. All these fisheries have changed and yet the catching of shellfish is still best achieved by use of the old and tested formulae.

37

Regulation on size Regulation of the crab and lobster fishery is controlled by a limit on the size of crustacean which can be landed. At one time the sale of berried hen lobsters was prohibited on conservation grounds, a ban which was lifted in 1966. Many fishermen who were involved in the trap fishery then and today consider that where fishing pressure is intense the landing of berried females is one cause of decline in stocks. When the ban was in operation many fisherman adhered to it and returned all berried females. At the same time there were also the unscrupulous who would 'scrub' the eggs from the hen producing a clean lobster.

Fig 13 Measuring a lobster from the eye socket to end of carapace

While technology exists to stain the egg carrying pleopods to see if the berries have been scrubbed or if the eggs have been shed naturally, scientists are divided in opinion as to whether the method is conclusive. At the moment it is unlikely that a successful prosecution could be brought on the evidence of this technique alone.

There can be little point in the reintroduction of the ban until legislation is tightened with the provision of a foolproof method available to fishery offices to test if hen lobsters have been scrubbed. If all berried hens were returned by all fishermen, while there would be a period when catches would fall, when the eggs had

Fig 14 A lobster which just measures the legal landing size

hatched and dispersed naturally, these hen lobsters would once again be legal quarry.

At the moment it is the carapace measurement which controls the fishery and market, a dimension taken from the eye socket to the edge of the shell on the carapace. This method of measurement is a much more precise one than the old measurement of overall body length, measured from the beak to the edge of the hard shell on the tail. This old system was open to abuse, it being possible to 'stretch' up to legal size a lobster which was slightly under. Nor was it unknown for the size to be brought up to the required measurement by 'extensions' from a lobster's foot fitted over the beak. In the present system there is no provision for such practices; what must be watched for is that a gauge made from thick metal will provide a different measurement than one of thinner metal.

Recent increases in the minimum legal size of lobsters have now brought about a situation which was unheard of a few years ago. Previously it was unknown for a lobster carrying eggs to be under minimum legal landing size, today they occur quite frequently. While there is no scientific evidence to back this up, it is the opinion of many fishermen that lobsters are breeding smaller and younger – a reaction perhaps against the fishing pressure upon stocks.

When the method of measurement changed from overall length to carapace length, this was set at 80 mm (3·15 in). Since then the size has been raised twice, initially to 83 mm (3·22 in) and then to 85 mm (3·34 in). What is interesting to note is that with the original 80 mm size some lobsters which failed to meet the size by the old mode of measurement were quite large enough on the carapace length. Likewise some under the carapace measurement would have been large enough by the overall length. This seems to indicate that as in humans there is a great difference in physical make-up between individual lobsters. Even with the present 85 mm carapace measurement,

39

sometimes when one side is measured and found to be undersize, the other will meet the legal requirements.

Further protection exists for the lobster in that it is illegal to land or offer for sale any parts of lobster unattached to the whole animal. This would specifically exclude the landing of lobster claws and tails in an attempt to by-pass the carapace measurement rule.

Until recently (4 April 1986) a uniform minimum landing size for crabs was in force throughout the UK. This has been changed and sizes now depend upon the area where landing takes place and the sex of the crab. The following are the sizes currently in force:

Area	Minimum size
South coast of England between the eastern boundary of East Sussex and the western boundary of Dorset	Male 140 mm (5·51 in) Female 140 mm (5·51 in)
Devon and Cornwall	Male 160 mm (6·3 in) Female 140 mm (5·51 in)
The coast of Wales from its southern border with England to Cemaes Head in Dyfed at 52° 07' north latitude	Male 130 mm (5·11 in) Female 130 mm (5·11 in)
The remainder of the British Isles	Male 115 mm (4·52 in) Female 115 mm (4·52 in)

Note There are proposals to amend these minimum sizes from 1st January 1990
National minimum   125 mm (4.93 in)
Somerset, Avon and Gloucestershire   130 mm (5.11 in)
Lincolnshire and Norfolk to remain at   115 mm (4.52 in)

Protective legislation This 1986 order also prohibits the landing of claws detached from edible crabs from within the British Fishery limits by British boats. This latter part of legislation was considered by many fishermen to be long overdue. When gluts of crabs were caught

40

merchants could not cope with the volume of brown meat from within the shell. This led to the request to fishermen to land the claws only containing the white meat which was in greater demand. Meanwhile the unfortunate creature was returned to the sea-bed to starve or grow new claws, whichever came quicker. In fairness it must be said that in many areas fishermen refused to land claws only. When they did remove claws the crabs were killed by spiking rather than returned to the sea crippled.

What the crab does enjoy is total protection of all females carrying spawn, the landing of egg bearing females being totally banned throughout the UK.

As yet there are no moves to regulate the crab and lobster fishery in the same way as that on the eastern American seaboard for the American lobster (*Homarus americanus*). Like the European lobster, this is a true member of the genus, not to be confused with more distant relatives such as the spiny lobster.

Some north American lobster fisheries are regulated by traps being provided with escape holes to allow any undersized shellfish to escape. Other measures on the western side of the Atlantic include some device incorporated in the trap to decay or destroy and allow any catch to escape in the event of the trap becoming lost and a 'ghost fisher'. It is doubtful if such a measure is needed in the UK. Rarely in this country do creels which have been lost or lain unattended for any length of time contain any notable numbers of shellfish. Other American regulations include close seasons and a limit on the number of creels or traps any boat is permitted to fish.

At the moment there is no immediate danger of the fishery for wild lobster suffering the same fate as that of the Atlantic salmon in Britain. Over the past ten years there has been a massive increase in the numbers of 'farmed' salmon on the European market. This continual year round supply has been to the severe detriment of the commercial salmon fishery. Prices for wild salmon have remained static or have in fact fallen in the face of the farmed fish.

Improvement in lobster habitat A further method of increasing lobster stocks is to improve the habitat on the sea-bed. A great deal of work has been done in America on this subject of creating artificial reefs to improve lobster catches. One conclusion which was reached was that a selection of half round concrete pipes provided as much habitat as a large complex and expensive structure. Any fisherman who contemplates a do-it-yourself operation in sea-bed improvement should be aware of the implications of the 'Dumping at Sea Act' which guards against pollution and any infringement in the rights of free navigation.

Market protection Classed as a luxury food, there are as yet no plans to provide market protection for lobsters within the EEC framework. Nor does the EEC appear to have any plans to protect the home catching industry against the dumping of imported Canadian lobsters on the European market. These Canadian lobsters frequently flood the markets within the EEC, after being flown across the Atlantic, at prices far below those which would be economically viable for unsubsidized home caught lobsters.

Moves are being made to provide the crab market with some kind of market support system. At the moment two schemes are proposed for holding crabs withdrawn from market if they have failed to reach a set minimum price. Of the two, cooked and frozen appears to be the more practical, as the alternative is for live storage in wire cages.

2 Vessels; finance, boats engaged in crab and lobster fishing, propulsion

If crab and lobster fishing is to be engaged on a full-time commercial basis some form of vessel is an essential requirement. While it is quite possible to take an astonishing number of summer lobsters from traps worked from rock edges, this could only be considered a part-time pocket money occupation. Full-time fishing without a boat is restricted by the amount of gear which any one person can carry. A strong back, sound wind and nimble legs are needed, as some of the most promising spots for fishing creels from rock edges are both remote and physically difficult of access.

The vessel   Several options are open to fishermen wishing to purchase a vessel for trap fishing. At one time it was a simple matter of having a few hundred pounds to invest in an ex ship's lifeboat hull and fit it out with a second hand diesel taxi engine and you were in business. Another attractive option would be to purchase a bare GRP (glass reinforced plastic) hull from a manufacturer for self completion, fitting out up to a reasonable standard is well within the scope of most people reasonably handy with basic tools.

Fitting out a hull can take a surprising amount of time – working part time a 23 foot hull such as supplied by Island Plastics might take up to twelve months. Concentrating full time and buying timber already sawn and dressed to size could cut this down to three or four months.

Fig 15 Fitting out a GRP hull is well within the capacity of most handymen

Finance By far the majority of fishermen will today opt to purchase a new or second-hand vessel ready for working the fishery.

Depending upon size a suitable vessel for trap fishing can cost anything up to £100,000, even more if equipped with vivier hold or chilling facilities. Few fishermen have this kind of finance readily available, making it necessary to obtain money in the form of grants before purchase may go ahead. For most established fishermen contemplating a new vessel the first approach for financial assistance is usually made to the Sea Fish Industry Authority, who took over the administration of such funds from the White Fish Authority and the Herring Industry Board.

Before any money is given the SFIA will require to see audited accounts from previous operations, plus a coherent set of proposals regarding operation of the new vessel. Directly funded by the Government the SFIA must ensure that the funds available are spent in

44

a proper manner. It is unlikely in the extreme that a new entry to the fishing industry would be considered for any form of assistance from the SFIA.

At the time of publication financial aid from the SFIA to build new vessels is restricted to a thirty per cent direct grant made at the Authority's discretion. It is the fisherman's responsibility to obtain the remainder of the purchase price.

Unfortunately, due to their smaller size, the majority of boats built for the crab and lobster fishery do not qualify for a further 25% grant from EEC funds. For boats over 12 metres (39·37 ft) the EEC grant may be applied for, but this remains something of a gamble as if this is qualified, payment is not forthcoming for a considerable time.

It is also possible to take over, in the purchase of a second hand vessel, any SFIA liability held by the previous owner. Once again this will mean negotiation with the Authority before this step can be taken.

Application can also be made to the SFIA for grants for boat improvements such as re-engining or fitting new deck machinery. As the standards of installation and safety are very high in work carried out with SFIA assistance, if the vessel is not already up to this specification, there may need to be additional expenditure to achieve these standards. This could involve fire fighting, bilge pumping and gas detection improvements for which a grant would also be available to approved applicants.

If a second hand boat does not involve the SFIA, the next most suitable source of money is through the branches of the local bank. Bank managers in fishing communities are usually well aware of the borrowing needs of the fishing industry, but like the SFIA will need to see a proper set of proposals concerning the repayment of any loan.

Vessel size and design In the trap fishery, more than in any other branch of the industry, will be found a wide diversity of size and

45

hull forms. Somewhere around the shores of these islands can be seen at any time, working at crab and lobster fishing, anything from 16-foot ex ship's lifeboats right up to the sophisticated 50-foot-plus vivier tanked offshore crabber. In between there will be purpose-built and multi-purpose vessels in a clinker or carvel wooden construction, GRP, steel, aluminium and even concrete.

As yet there is little information available about the prospect of lobster fishing on the edge of the continental shelf as is found in North America. This fishery is pursued with ocean going vessels working on a 'trip' basis with the catch measured in thousands of pounds in weight. Here are caught, in huge traps requiring machinery to handle them on deck, lobsters which most fishermen in Britain can hardly even expect to dream about. Whether a similar opportunity exists on our side of the Atlantic is pure conjecture. The undertaking would certainly be beyond the scope and range of any of the boats currently engaged in trap fishing in the UK.

By far the majority of the boats working from low-water mark to five or six miles offshore are around 25 to 35 feet in length. Hull forms range between the deep beamy round bilge heavy displacement craft to the ultra fast triple hulled modern 'dories'. Different hulls all have their adherents, but where the planing type of hull will be restricted to static gear fishing, boats of heavy displacement can turn to some form of towed or mobile gear when trap fishing is slack.

Size influenced by harbour depth   Boat design was much influenced by the facilities in the home port of the vessel. Where all tide harbours are to be found much deeper drafted craft are built than where there are tidal restrictions. Also several forms of boat suitable for working from open beaches are in use, the most familiar being the coble of the northeast and the Cromer beach boats.

Fig 16 Fully decked boats such as the inside vessel are less popular now for crab and lobster fishing

Fig 17 Small tidal harbours such as Cove in Berwickshire are best served by small boats with a shallow draught

Fig 18 The 'coble' is still popular in Northeast England

A boat's creel fishing requirements

The main requirements of a boat for creel fishing is that it should provide a stable working platform and be capable of carrying enough gear to engage the fishery. Much of the operation of the creel fishery, both in hauling and shooting is carried out at slow speeds, leaving the boat very much at the mercy of wind and tide. Boats least likely to be affected by these influences will have a low freeboard; deep, long keel; and have the minimum of top hamper in the way of wheelhouses and shelter decks.

Position of wheelhouses

Traditionally, wheelhouses of fishing vessels were sited at the after end, a hangover from the days when steering was done directly by tiller. The aft wheelhouse is still favoured in many Scottish vessels, but there is an increasing trend to site the wheelhouse in smaller boats such as are engaged in the trap fishing sector in the forward position. This gives a clear after deck for gear stowage and handling but can lead to problems of siting the hauler far enough forward so that the boat will 'lead' along the gear when hauling. Forward wheelhouses do have the advantage of providing protection for the crew on deck and dual controls are much easier to arrange. To allow more forward hauling position some wheelhouses are offset to port or starboard. When fitted out in this manner it is often possible to avoid the need for dual controls, engine revolutions and steering being adjusted through a side wheelhouse door at the usual helmsman's position.

Even in boats of around 20 feet or under some form of wheelhouse or shelter is now considered essential. Vessels even smaller than this may carry at the minimum a VHF radio and echo sounder, neither of which stand being deluged in salt water spray.

Fig 19 Some fishermen prefer a forward, some an aft wheelhouse

On-board accommodation

While many of the craft engaged in the trap fishery are day boats, returning to their home port daily, in some parts of the country longer trips are undertaken. In the English Channel boats will work offshore for two or more days when weather permits. Around the Western Isles of Scotland vessels may be away from home for up to a week at a time, storing their catch in either keep pots in sheltered sea lochs or in vivier tanks on board. Such boats require full accommodation for the crew including facilities for cooking meals and comfortable sleeping berths.

Wooden hulls need protection

Boat hulls constructed from wood and GRP need protection from the constant procession of creels and anchors up her side in the way of the hauler. Continual

49

hauling will pick and abraid at these hull materials, eventually destroying them. It is usual to protect these areas with some form of cladding which is either resistant to wear or is more easily replaced than the hull material. In GRP boats this is mostly in the form of extra layers of glass fibre matting layed up over the finished hull. Wooden boats cause an extra problem with the possibility of moisture sealed in below the cladding, causing rot in the hull planking. In wooden boats the hull below any cladding must be well treated with timber preservative to prevent this occurring.

Many fully decked low gunwaled boats are still working in the trap fishery, as are smaller craft with non-watertight decks where any water coming on board must be eventually pumped from the bilges. The modern trend is for the boat to be 'well' decked and provided with self freeing draining ports. A convenient height or rather depth of 'well' deck is somewhere between knee and thigh height, allowing easy access over the side for lifting the gear on board. Three standard fish boxes high is about right, which also allows creels to be stacked two high within the depth of the 'well'.

Preventing loss of stored creels
Creels stowed on board a rolling vessel will, with every dip of the rail, endeavour to pitch overboard. If utter chaos is not going to reign, steps to prevent this are a necessity, most easily attained by fitting a raised rail in the stowage area. Raised rails greatly increase the carrying capacity and can be even further enhanced by a cage arrangement extending aft over the stern.

The engine
Thankfully the days when oar and sail were relied upon for propulsion are long past and unlamented. Today, at least in Britain, all full-time fishing boats make use of some form of mechanical propulsion to take them to and from the fishing grounds. Engine size and specification depend very much upon hull form and whether or not the boat is going to be employed in

50

any fishing other than with static gear. An engine of 60 horse power providing efficient propulsion for a 25-foot displacement craft will be totally inadequate in realizing the full potential of a planing hull of similar length.

Engine size should be selected to give economical cruising speed at 200 or 300 rpm below maximum engine rpm. Running at this slightly slower speed means less wear upon engine components and some power in reserve available to tackle adverse conditions.

In the modern situation marine propulsion means two things: inboard and diesel. Today there are available a wide range of diesel-fuelled engines to suit vessels of all sizes and hull types.

Boat engines are available in two forms, the proper purpose built marine units, slow revving and high torqued; or those capable of powering static gear fishing vessels, which can be one of the many marinized units based upon van and lorry engines. These are purchased from the engine manufacturers by firms specializing in converting or marinizing them with a number of bolt-on parts.

Marinization consists of fitting indirect cooling systems where the excess engine heat is dissipated through pumped sea-water rather than the radiator and fan in the original engine.

A complete marinization kit would consist of combined header tank/heat exchanger with the raw sea-water being drawn through an oil cooler by an extra pump, passing through the heat exchanger before being discharged overboard.

While the purpose built marine engine may have a longer life than a marinized unit, due to lower production runs and high specification, they can also be twice as expensive.

Spares for the marinized engines are in general more easily obtained and cheaper than those for a marine unit. In most instances basic spares for a marinized unit are available – if not from the local garage, the nearest authorized dealer of the parent engine will have them available. While a network of dealers back up the sales and servicing of *bona-fide* marine engines, they are not nearly as widespread as outlets for marinized units.

While most marine engines are now cooled by indirect water circulation there are still a number of engines available with air cooling. These are best suited for open boats where the supply of a large volume of air to the engine fins is easily arranged. Where air-cooled engines are below decks, adequate ducts and vents for the intake and discharge of the air must be arranged.

Adapting an engine for marine work

Self marinization of a basic engine has been the means of many aspiring fishermen obtaining their first boat. First the engine itself must be sound; if there is any doubt it should be stripped down and all remedial work carried out. With the engine rebuilt it is quite a simple matter to bolt all the necessary parts on to provide a fully marinized unit.

Methods of transferring power

With an inboard engine installation there are several methods of transferring the power to drive the boat forward. By far the majority of boats attain this with a reduction and reverse gearbox driving through a rigid shaft to the propeller. Few boats today are fitted with mechanical gearboxes; for convenience and ease of operation the oil operated, or hydraulic gearbox is unparalleled. In these boxes oil or hydraulic power does not transfer the engine power to the shaft, rather it is the pressure of oil moving a control valve giving the selection of ahead, neutral and astern at the flick of a switch. This removes the physical effort needed to engage and disengage gears and lends itself well to remote control through flexible cables.

Not all vessels transfer power via a rigid shaft arrangement. An alternative is the inboard/outboard drive, a method often selected for boats with fast or planing hulls. Only boats with transom sterns are really suitable for this arrangement with the engine mounted aft hard against the transom. The outdrive unit itself resembles the lower half of an outboard engine but is driven from inboard. Steering with this set up is as with an outboard with the entire propeller units turning to port or starboard. A useful feature of this arrangement is that the propeller can be hinged upboards to clear fouled gear.

For boats working any kind of static gear, the jet drive unit would appear to be best. In this installation motivation is obtained via a high pressure pump, working on the same principle as a jet engine with a rearwards thrust providing forward motion. Steering is by deflection of the thrust to port or starboard and provision is made by a scoop arrangement for going astern. Like the inboard/outdrive, jet units are most likely to be found on high speed hulls.

A further option remains in hydraulic drive; in this the engine drives a heavy duty hydraulic pump which in turn is connected by piping to a motor at the inboard stern gland position. No gearboxes are involved and the engine can be sited in what is the most convenient part of the boat as there is no shaft to fit and align.

Although these last three options are popular in some of the pleasure boat markets, by far the majority of working boats will be fitted with the traditional shaft and stern tube arrangement.

Outboard motors   On board smaller boats the expense of installing an inboard engine may not be strictly necessary. Modern outboards are reliable pieces of machinery compared to those of the past. Now available in four stroke and diesel-fuelled versions, the plug oiling purgatory of the two-stroke oil and petrol mixture need no longer be endured. Installing an outboard is a simple matter of first ensuring that the transom or mounting bracket is strong enough to take the thrust and clamping the unit

53

to the boat. Where theft is a possibility, some device is required to prevent it being removed for it is a simple matter for the unauthorized to unclamp it and depart with it.

Nor need the choice of an outboard condemn the fisherman to the tedious business of hand hauling his gear. Some modern outboards are provided with a stub shaft from which power can be taken to drive a self hauling vee wheel.

Whatever type of propulsion unit is chosen the engine should be well maintained. As is mentioned elsewhere in this book, a great deal of lobster fishing takes place close inshore where there is little sea room. It is unnerving to be caught trying to haul traps close inshore with an engine which is showing temperamental tendancies, or worse still has stopped completely.

Diesel engines Many of the faults in diesel engines relate directly to the fuel supply, or rather the lack of it. Filters should be changed at regular intervals in case they become choked with sediment from the fuel. Where fitted water traps need regular checking and the water drained; when the vessel has been inoperative for some time water and sediment will drain to the lowest part of the tank. If drain cocks are provided these should also be opened to release the detritus into the bilges.

To provide ignition within the cylinders, the diesel engine depends upon an exact supply of fuel at a designated pressure. Three components are involved in providing this in the installation. The fuel pump raises the fuel from the tank to a level where the injector pump takes over, forcing it through the injectors in a precisely metered quantity.

Maintenance At least once per annum the lift, fuel, pump and injectors should be removed from the engine and taken to a specialist for cleaning and recalibration. This job requires special equipment and knowledge and is well worth the annual cost, often returned in fuel economy if not peace of mind.

The other most likely cause of engine failure is overheating, mostly due to the failure of the raw or sea-water supply to the heat exchanger. When working in weedy or dirty water the sea cock strainer should be checked daily, although this does not prevent a loose sheet of plastic being drawn against it at the most inconvenient moment.

For long engine life and minimum wear on moving parts the lubricating oil must be changed at intervals recommended by the engine maker. At the same time the filter should be renewed, with the modern cartridge oil filter, washing out with diesel oil is a false economy.

It is interesting to note that when small boats were powered by petrol/paraffin engines, many lobster fishermen in the northeast of Scotland ran them entirely on expensive petrol purely for reasons of reliability with the 'cleaner' fuel.

3 Hauling equipment and other auxiliary controls

In addition to the driving machinery, the trap fishing boat needs further equipment to retrieve the gear. With the economic need to work at the very minimum of a hundred creels per crew man, it would be a backbreaking if not impossible task to hand haul them when rigged in fleets.

Selection of hauler   For serious full-time fishing some form of hauler is needed to bring the creels from the sea-bed to the surface. Before selecting the torque of a hauler it must be borne in mind that when working gear in fleets or tiers, it is not so much the creels which are being hauled to the boat as the boat along the line of creels. Also, what is of equal importance, a hauling device which is too powerful will result in broken rope and smashed creels when working over hard ground. The best idea is to fit a hauling package of more than adequate power, but provide it with a pressure relief valve which will bypass the oil from the hauler motor when extreme weight comes on it.

Trap hauling   Originally devices for trap hauling were powered from
machinery   the front end of the engine through a series of belts, shafts and gearboxes to the hauler head. Drive for

56

these was controlled by either a dog clutch or jockey pulley, but lacked the facility of knocking easily out of gear in an emergency.

In some areas of the country an auxiliary engine was provided to power deck machinery. This practice is less so today, with most boats taking their drive to deck machinery from the main engine except in cases where in some of the larger boats a second engine is fitted to provide extra electrical power or drive a water chilling system for vivier tanks.

Not all trap hauling machinery is dependent on its power from main or auxiliary engines. A compact self-contained unit suitable for small boats is available from 'Seawinch' of Bridport in Dorset. Weighing only 80 lbs with a pull of 400 lbs at 200 ft per minute the hauler package can be sited on board in the most convenient position.

Fig 20 Capstan type hauler

The advent of reliable compact marine hydraulics has made most other hauler installations redundant. Two types of hydraulic systems are in existence which are suitable for powering deck machinery on small vessels (basically the only difference is in the type of pump used to power the systems): the variable delivery pump which runs constantly and does not require to be

57

separated from the engine by a clutch, the oil being diverted automatically when the hauler is not in use; alternatively, the set-up will be with a constant delivery pump, clutch controlled from the fore end of the engine. Constant delivery pumps should not be run at high revs when the boat is steaming, or damage to the internal components will result.

Drive from the engine can be through pulleys and belts to the pump, a system where changing pulley sizes to vary revolutions to the pump can be used to regulate the final speed at the hauler head. The system is prone to troublesome slipping belts and can cause wear on the pump bearings due to the sideways pull.

Coupling the pump directly in line to the engine is the most reliable method with there being no belts to stretch or break. When this installation is used there is no way to adjust the final speed at the hauler except through changing the pump capacity or by altering engine revolutions when the hauler is engaged.

Hydraulic systems enjoy the facility of using the control lever to instantly stop or reverse as the situation requires. Pipes to the hauler head can be led in any direction to a suitable position, plus the fact that it is convenient to break into the system at any point to provide power for any other deck machinery.

Fig 21 Self hauler patented by Robert Hope of *North Wind BK 33.* When the rope became tight on a fastener the spring is compressed so helping to avoid damage to gear

58

Capstan heads were at one time the standard fitting for creel hauling right to the days when hydraulics were becoming popular. Like the flat plate hauler, these are rapidly becoming an anachronism as more and more boats convert to self-hauling vee wheels. Capstan and flat plate haulers need the constant attention of a crewman to pull the rope away behind the hauler head. While the capstan takes most of the weight, operating it is a monotonous job with the turns of rope needing to be taken from the head as each creel is lifted on board.

Therefore, most fishermen today opt for a hauler which can work unattended. The principle of the self-hauling vee wheel is a simple one, consisting of two convex discs bolted together over a central shaft. As the wheel turns, the rope, weighted by the gear, is drawn inboard. This it does until it comes up against the knife or stripper blade, which ejects the rope from the wheel throwing it down in loose coils on the deck.

Haulers may be mounted in the vertical or horizontal plane either on a 'P' bracket hauling directly on board, or through an outboard swivelling davit arm. It is important that the wheel fits the rope being used, adjustment being made by adding or removing shims from the centre shaft. Eventually through use a rim will wear on the inner lips of the hauler. This prevents the rope from jamming into the wheel, and as the boat

Fig 22 Combine 'Vee' wheel and capstan hauler; in this case the rope is led through the roller fairlead

59

rolls, rope will constantly be slipping and springing free.

Skimming this rim off – a job best done on a lathe by skilled labour – will restore the rope gripping properties. Most wheels will stand this treatment up to three times before the metal becomes too thin, when the only course remaining is replacement.

When the wheels are set up correctly the hauler can work entirely unattended with no need to stop it when lifting strops and creels inboard. That they are now the universal choice of most fishermen is obvious, as it leaves a crew member free to carry out other tasks.

With a 'P' bracket hauler mounted close to the gunwale there is no need for any roller or any other device to guide the rope on board. Davits have an open ended hanging block on the end of the arm, preventing wear and friction on the rope. Where a capstan head is used, or a vee wheel is mounted in a horizontal plane, a roller fairlead, jeanie or mollologger must be rail mounted to lead the rope from outboard to the hauler.

Fig 23 'P' bracket hauler with self hauling 'Vee' wheel

Position of controls Despite self haulers having taken much of the drudgery and effort out of hauling, it is still convenient at times to be able to steam the boat ahead and adjust steering from the working position on deck. To free a man from being constantly in the wheelhouse, it is essential for dual controls to be mounted as near to the hauler position as possible. Most engine controls are now operated through flexible cables instead of the less convenient rigid rod arrangements as in the past. For ease of operation it is best if the two functions of engine revolutions and gearbox control are carried out through separate levers. The twin lever set-up is less cumbersome to operate, but there must be an inbuilt safeguard within the system to prevent the gearbox being engaged with excessive revs on the engine.

As with engine and gearbox control, steering on deck at the hauling position is a convenient affair. In this respect the hydraulic systems are unbeatable. Hydraulic steering has the advantage of keeping the rudder in the set position when the boat is steamed ahead or rolls in rough weather. A rudder indicator which can be seen easily from the deck will immediately show the helmsman the set on the rudder.

Fig 24 Alternative layouts for haulers and stowage on boats with aft wheelhouse

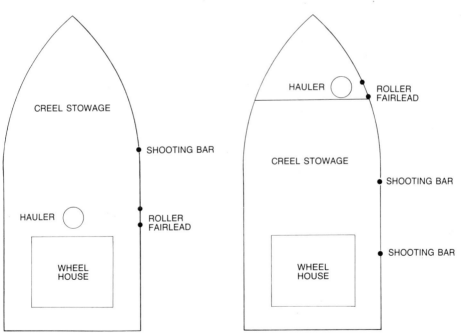

Safety of stowed creels Presuming that the serviced creels are stowed aft on the vessel, a means of preventing the main rope being drawn over the stern of the boat when shooting must be used. Shooting bars or irons set into the gunwale are the most common means of achieving this. On board larger vessels with long fleets of gear, two might be needed, removing the forward and working from the after as shooting progresses. Using two shooting bars reduces the amount of walking which the crewman needs to make from stowed gear to shooting point. In fact, by moving to another shooting bar, it is the shooting position which is being altered bringing it nearer to the stowed creels.

Advantage of a mizzen sail One old fashioned item which is still appreciated today by all fishermen who work static gear is the mizzen sail. With the sail set and careful manipulation of the rudder to counter tide and wind, hauling up through

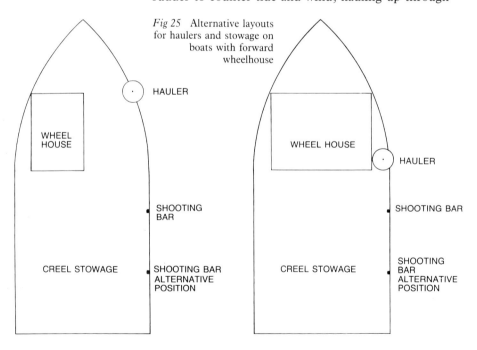

Fig 25 Alternative layouts for haulers and stowage on boats with forward wheelhouse

HAULER

WHEEL HOUSE

WHEEL HOUSE

HAULER

SHOOTING BAR

SHOOTING BAR

CREEL STOWAGE

SHOOTING BAR ALTERNATIVE POSITION

CREEL STOWAGE

SHOOTING BAR ALTERNATIVE POSITION

even winds up to force seven can be accomplished with an occasional kick ahead on the engine. A hazard does exist when shooting out gear before the wind with mizzen set. Should the rope foul a crewman the boat can be blown beam on to the wind and draw tight before the bow can be brought round. For the safety of those working on deck the greatest care must be exercised during these circumstances.

Cargo derrick Where crab catches are heavy it is worth while fitting a cargo derrick for swinging the catch ashore – a useful item if other forms of fishing are to be pursued. If the catch is stored in heavy keep pots at sea a landing derrick for bringing those inboard will be required. For operating these derricks the self hauling head is not a practical proposition and if no separate power is at hand a combined vee wheel and capstan head is the best choice.

Avoidance of propeller fouling Any kind of fishing gear is prone at some time or another to become entangled on the propeller – serious business at any time, but doubly dangerous in inshore waters. At one time this meant that the boat would be disabled if the obstruction could not be freed from on board by the crew. Some boats were fitted with inspection hatches or 'tunnels' to the propeller. In the poor weather conditions which usually lead to prop fouling, the inspection hatch could often do little more than confirm that the prop was in fact fouled.

Today there is available a simple and efficient device which should, providing no actual creels are in the prop, make disablement through prop fouling a thing of the past. This consists of sharp stainless steel blades mounted on the stern tube pointing towards the prop. With these fitted one kick ahead and astern should cut free even the thickest rope.

63

4 Electronics, sounders, radar position finding and radio

The location of prime shellfish ground is something which is handed down through generations of fishermen, experience usually being gained by working aboard boats already engaged in the fishery. Most of the larger areas of hard ground suitable for shellfish are to be found marked on charts of the area being worked. Charts do not tell the whole story and grounds named upon them may be called something entirely different by local fishermen. Small patches of hard exist in areas where the chart soundings do not indicate any; often these are very productive, especially for a good class of lobster, as they may not be so exploited as larger areas.

Many now well known grounds were originally discovered by accident – sometimes it was after gear had been shot in fog. Before the days of echo sounders there was no practical method of knowing what was below the keel when shooting. Hauling the following day in clear weather would produce catches from grounds which previously had been thought to be 'away off the marks'.

In a similar manner at the onset of a blow, gear might have been moved offshore into what was thought the protection of deep water and soft ground. Only when hauling and finding that the gear had been destroyed by hard ground were some of the offshore banks and reefs located.

The echo sounder Today the old methods of towing a weight or working an armed lead are seldom used in the search for new shellfish grounds. With the advent of accurate, compact marine electronics, suitable for even the smallest of vessels, the nature of the sea-bed is instantly available at the touch of a switch. More than any other instrument, it is the echo sounder which is the principal electronic aid to the modern trap fisherman.

Sounders all work on the same basic principle of signals transmitted to and from the sea-bed. These signals bounce from the bottom and are transformed electronically to be displayed on board in a recognizable representation of what lies below the boat's keel.

Sounders are of high or low frequency, the exact mode usually being determined by the type of transducer used. This is the sender and receiver unit normally fixed through the boat's hull in a position where turbulence from forward motion or propeller will not cause false signals. High frequency sounders are mainly used for the determination of bottom-swimming fish shoals.

For the shell fisherman it is the composition of the sea-bed which is of the greater importance; here the need is best served by a sounder of a low frequency. It is possible with a sounder of a dual frequency to enjoy both these facilities, changing over if another kind of fishing is employed.

Depth and ground recorders Depth- and ground-recording machines can display their findings in several different modes. This can be anything from a simple flashing neon on a circular calibrated dial showing depth and giving some indication of the bottom conditions, to (the other end of the price scale) ultra-sensitive colour video sets, displaying their findings on a colour television screen. When correctly tuned in, colour sets show the slightest change in the nature of the ground immediately the boat passes over it. In between the neon and colour sets there are black and white video, and the chart recording types. In the latter a stylus traces the contours on sensitized paper moving through the machine. Varying thickness and intensity of the drawn line gives the operator an interpretation of the depth and condition of the sea-bed below the boat. Chart recording sounders are useful for keeping a permanent record of the ground, a facility not readily available in the video sounders.

These are the most common types of sounders in use in the shellfishing industry. There are also audio sounders which 'bleep', becoming faster when hard ground is crossed. At one time there was even a variant on this theme in a sounder which 'talked' but this never achieved any great popularity.

Video sounders Video sounders have the merit of costing nothing to run and can be left switched on in the hope of locating new ground. While the cost of recording paper for chart sounders is relatively cheap compared to the price of lobsters, most fishermen are reluctant to leave them running constantly. An adequate chart recording sounder can be purchased for between a third and a quarter of the price of a colour video set – the difference can buy an enormous amount of recording paper.

Chart recording On chart recording sounders a paper width sufficient to
sounders show a double echo will often reveal patches of ground and weed not shown on the top mark. In most modern machines the facility to expand the bottom can give an enlarged picture of the nature of the ground below the keel. The sounder of today can be set to track the bottom, retaining an expanded picture of the sea-bed

66

regardless of the depth, which is displayed in a digital form on the set.

Fig 26 Examples of hard weedy ground as shown on graph-recording echo sounds

A Typical rocky, weedy ground favoured by lobsters

A

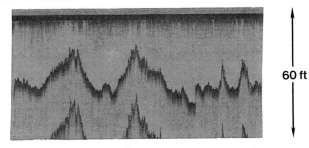

60 ft

B Over ground like this creels shot too tightly could be suspended in mid-water

B

60 ft

C This is similar to the reef shown in *Fig 1* with X and Y being suitable for a good class of lobster

C

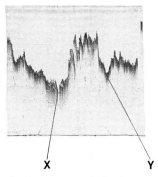

X   Y

Navigation systems Working close inshore around the low water mark, small boats will be steered visually around outcrops, boulders and weed beds when placing the gear. Farther offshore where good landmarks are available, the creels will be shot, with the aid of the sounder, according to

67

land based transits or 'marks'. These methods work in clear weather when within sight of land; vessels working farther offshore need additional aid in finding ground and picking up dahns. Offshore, the most used aid for many years has been the Decca Navigator system, based on radio signals from shore stations providing cross bearings. These signals are picked up by the wheelhouse set and interpreted into 'lanes' deciphered by special charts showing them.

Fig 27 Boats are steered visually around the rocks inshore without any navigational aids when shooting creels

Lately several other receiving sets have come on the market which 'pirate' the signals provided by the Decca company. Computers inside these receivers translate the Decca 'lanes' into grid reference of latitude and longitude as shown on the normal chart. While the Decca receiver is only available on annual hire, the pirate sets can be purchased outright making the use of radio based navigation available to boats which would normally not afford to use it. Financial

68

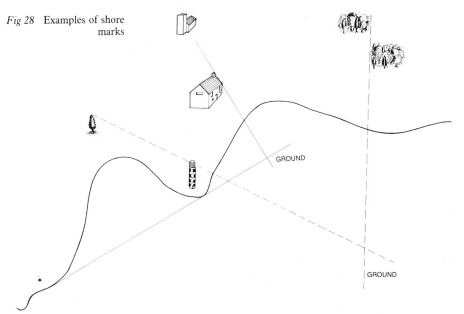

Fig 28 Examples of shore marks

GROUND

GROUND

responsibility for the transmission of the Decca signals serving UK waters has now been taken over by the British Government. This leaves fishermen free to choose the signal receiver most suited to their operation and pocket, either on an outright purchase or rental basis.

When the receiver for the Decca signals is set up correctly and transmitting stations give accurate signals, position fixing can be accurate to within a few feet. The hours of darkness will affect the signal received as do different weather conditions, snow and rain being likely to cause deviations in the position indicated.

Note Mk 21 Decca receivers were sold to skippers who already had rental contracts when Decca signals were taken over by the Government.

New Mk 53 Decca receivers are available of either rental or outright purchase with maintenance contract.

69

Fig 29 Decca navigator in
background, radar in
foreground

Fig 30 Three point radar
position fix using distance
and bearings

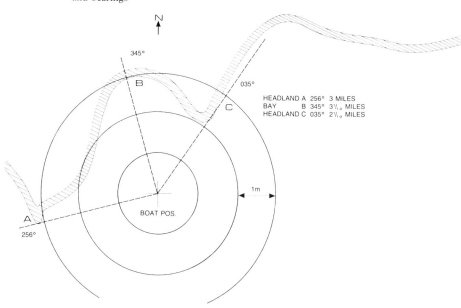

HEADLAND A 256° 3 MILES
BAY   B 345° 3¹/₁₀ MILES
HEADLAND C 035° 2¹/₁₀ MILES

70

The up-to-date radar is a useful aid to working any form of static gear marked with dahns. On a modern set with variable range and bearing markers a very precise fix can be arrived at when working within reasonable distance of shore targets. While not as accurate as radio bearings it is possible to get close enough to the dahns to pick them up with the set switched to short range. Of course, the use of radar is also used as a means of safe navigation and collision avoidance in conditions of poor visibility.

Few boats working solely at crab and lobster fishing can justify the expense of sonar. Like sounders, the sonar depends on an underwater signal transmitted and bounced back from the sea-bed. Unlike the sounder

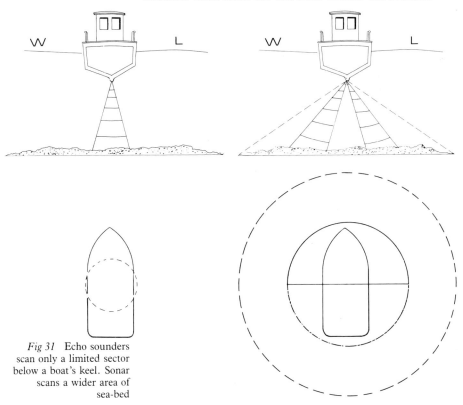

Fig 31 Echo sounders scan only a limited sector below a boat's keel. Sonar scans a wider area of sea-bed

which only shows the ground below the boat's keel, the sonar scans with a revolving transducer. Sonar can in fact be likened to an underwater radar giving a picture of what is happening below and around the boat. Sector scanning is a feature of the more upmarket sets, which have the facility of being able to search in specific arcs ahead and on either beam of the vessel. But for the price, sonars would be a useful tool when shooting on odd patches of hard with the helmsman steering the boat according to the sonar signals.

Few fishing boats today operate without the aid of VHF radio. In the trap fishery it is almost an essential aid when working on heavily fished grounds. The ability to call up another vessel and enquire in which direction gear is shot prevents (or at least it should prevent) the chaos of gear being shot across that of another boat. VHF is also a useful communication channel ashore, both to the coastguard services and salesmen and the boat's agents. With a private channel ashore to the agent's office, pressing matters such as ordering bait supplies and ascertaining markets are immediately available.

5 Creels, pots, their treatment and materials used in construction

Today around the shores of Britain there are two basic styles of shellfish trap in regular use. These are the inkwell design of the English Channel and the southwest, and the creel shape of the east coast and Scotland. At one time the French style barrel was also fished on the Channel grounds but due to their vulnerability are now little used. Although the shape of inkwell pots remains similar, with creels there are subtle differences in design, even between neighbouring ports. Each community likes to think that their design is superior to all others in catching and holding qualities.

Local conditions determine type to use

This may not be so fanciful as it may first seem. Trap design has evolved over a period of many years to suit local conditions. This may be the nature of the ground, depth of water and strength of tide. Creels used in the strong tides of the Pentland Firth between the Orkney Islands and mainland Scotland, for instance, will require twice the weight to hold them on the sea-bed than on the east coast. Many fishermen have tried the inkwell pot on the east of Scotland and found it lacking; no doubt the east coast creels have a similar reputation in Cornwall. Yet on some of the outer isles the inkwell pot is gaining an increasingly good reputation. Perhaps conditions are similar there to the grounds they were designed for.

Experimental fishing conducted by Marine Laboratory, Aberdeen (Shelton and Hall[7]) to compare the efficiency of pot versus creel were carried out to the west of the Outer Hebrides in September 1977. Conclusions reached were that there was no difference in the number of crabs and lobsters taken by the two types of gear. Due to the larger eye or entrance, the inkwell pot caught larger crabs than the creel; there was little or no difference in lobster sizes between the two types of gear.

The gear used in the experiments consisted of steel inkwell pots, 63 cm (24·8 in) diameter at base, 40 cm (15·7 in) at top, 45 cm (17·7 in) high with rigid plastic eye of 20·3 cm (8 in). Creels were of semi traditional construction with 70 cm × 46 cm (27·5 × 18 in) wooden base, galvanized steel hoops or boughs, with hardwood sidesticks. Two eyes were incorporated in the creel, both of hard plastic and 12 cm (4·7 in) in diameter the finished creel standing some 31 cm (12·2 in) high.

Full results of these experiments are published in 'A Comparison of the Efficiency of the Scottish Creel and the Inkwell Pot in the Capture of Crabs and Lobsters' by R G J Shelton and W B Hall.[7]

Norfolk creel On the Norfolk coast at Cromer the creel is constructed in an entirely different way from that in north Northumberland. In the Norfolk creel there are four semi-circular hoops or boughs rising from the base to support the netting cover. In this creel part of the base is formed by a heavy cast iron fitting and the netting is fitted inside the framework. Eyes or entrances are fitted or worked in a different method from that used in other regions. Here the eye is situated in the centre of the middle panel, with the netting carried through on to the top of the opposite

74

side eye. Only the bottom of the eye is left unknitted, allowing the catch to drop down into the creel.

Northumberland and Scottish creels

Northumberland and Scottish creels are formed on a wooden base with only three boughs and generally have the eyes at diagonally opposing sides with the netting cover outside the framework. These are the basic differences, but method of constructing the base and eyes vary from region to region. Some will favour two hard eyes, others one hard and one soft or even two soft eyes. Doors for removing the catch and baiting seem to be another matter on which fishermen seem to be able to reach agreement as to which is the best method.

Inkwell pots

The inkwell pot shows less variation except in dimensions and materials. Inkwells all follow the same design as was originally used in the soft withy pots. Today rather than the easily destroyed willow wands, this trap will more likely be made from steel and plastic or a combination of both materials. Inkwell pots all have their necks or funnels sited centrally on the top, these mostly being preformed from plastic ready to clip into the pot.

A hybrid pot is used in some localities: this has a top entrance like the inkwell pot but is shaped like a creel. It would appear that the main advantage of this design is that it is possible to have a top entrance trap and still have provision for fitting a catch retaining parlour.

To make or buy?

A strong tradition remains in some regions for creels to be made entirely by fishermen themselves. In other places where fishing is more prosperous and catch prices firmer, pots and creels will be purchased complete and ready for the sea. There is much to commend the latter practice, especially where trap fishing is a viable proposition for the entire year. Buying creels ready to fish frees the fisherman from the drudgery and labour intensive task of constructing the

Fig 32 Spare time is spent making gear

ever-increasing amount of gear required to make a living.

Many firms now offer creels and pots for sale in various stages of completion. The most popular are bare frames of 8 (0·31 m) and 10 mm (0·39 in) round steel bar which are either galvanized or plastic coated or even given both treatments. Selection of suitable steel is most important in trap construction, with low carbon steel being much more durable than the high carbon content rod such as is used in steel reinforcement bars.

In addition to the bare frames, traps may be bought already covered with netting with all ancillary work carried out. Ten millimetre bar is a strong robust material for frame construction; unfortunately it also has the ability to chafe rapidly through the netting cover, even with the normal handling on board of hauling and shooting, without any damage being caused by heavy weather.

Use of steel and plastic

Steel traps, be they of creel or pot shape, are best bound with rubber strip both to protect the plastic coating and preserve the netting covers.

Plastic is probably the material of the twentieth century. The fishing industry as a whole has not been slow to realize the advantages of a material which is impervious to the destructive organisms which attack all natural materials when they are immersed in sea water. Plastic has a wide use in the modern fishing industry from the actual hulls of GRP boats, through twines, floats and buoys plus numerous other fittings on board fishing vessels.

The shellfish industry has been equally quick to seize upon the attributes of plastic and specialist firms now offer traps in either injection moulded plastic or of a welded plastic piping construction. Some are designed as basic frames to be covered with netting in traditional manner, while others incorporate the meshing frame and entrance as part of the injection moulding.

76

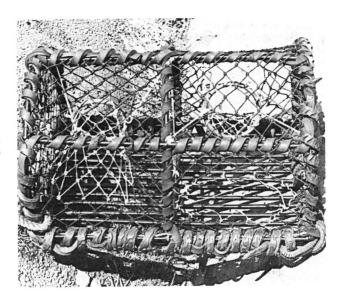

Fig 33 Steel creel bound with rubber tyre strips

Also on offer are creels which fold flat for stowage onboard the boat. At first these would seem to offer great potential with the carrying capacity of the boat greatly increased. Folding creels have yet to become popular with full-time fishermen, but further developments along these lines should not be ignored.

Fig 34 Left to right, steel base with solid plastic frame, plastic base and frame, tall plastic moulded pot

Methods of construction Where fishermen still construct their own traps, which is mainly on the east coast, the method of construction and choice of basic materials varies from district to district.

In the east coast creel, best described as of the rat cage shape of flat base, with a semi-circular frame covered in netting both, hard and soft woods are used in the construction. In southeast Scotland the base is formed from 'bulls' measuring 760 mm (30 in) long by 65–75 mm ($2\frac{1}{2}$–3 in) broad and 30–40 mm ($1\frac{1}{4}$–$1\frac{1}{2}$ in) thick. Due to the constant wear and abrasion on the sea-bed, a native hardwood is usually chosen to give strength and durability. In this respect oak is the superior timber, more expensive than beech, elm or sycamore, but well worth the extra cost when it comes to resisting wear. For best results, any timber used in creel construction should have been seasoned for at least a year before being put in the sea.

While the bulls form the longitudinal runners of the base, the framework is completed with transverse slats. Seven slats 460 mm (18 in) by 90 mm ($3\frac{1}{2}$ in) by 12 mm ($\frac{1}{2}$ in) are required for the standard base 760 mm (30 in) long. Slats may be either hard or soft wood, both having advantages and drawbacks. As can be expected, hardwood is less susceptible to the ravages of rot and marine worm but is prone to splitting when nailing, the reverse being true for softwood slats.

An alternative method of constructing the bases is used on the Yorkshire and Fife coasts, with the bases being formed from heavy softwood slats running in the longitudinal direction, with thicker broader transverse members at either end and centre.

With the base completed, the remainder of the framework is formed from two parts, boughs and sidesticks. These form the half round part of the frame carrying the netting cover and supporting the entrances and doors. Boughs are firmly wedged into holes drilled in the bulls, the frame being completed with sidesticks nailed and lashed in the longitudinal direction. On the

78

standard creel there are three boughs and sidesticks.

Traditionally, locally grown 'suckers' or saplings were used for the frame, usually being of ash, hazel or wild briar. While these made satisfactory creels for one season's fishing, on removal from the water they would dry out and become extremely brittle and liable to breakage when reimmersed.

Today the sapling framework has been changed to bamboo canes measuring from 19 mm ($\frac{3}{4}$ in) to 22 mm ($\frac{7}{8}$ in) in diameter, an import from the far east. A tough outer skin protects the soft inner core from the ravaging they must endure on the sea-bed. Notwithstanding damage by heavy seas, a cane framework should have the same two- to three-year life of a hardwood base.

An alternative framework can be constructed from 12 mm ($\frac{1}{2}$ in) internal diameter plastic water piping; a very resilient, durable and rotproof material. Frames are simple and easy to make with water pipe the boughs and sidesticks being bolted together with 6 mm ($\frac{1}{4}$ in) roofing bolts. A piping framework outlasts several hardwood bases with the only maintenance required being the replacement of the bolts. Reusable several times, these frames should outlast two or three hardwood bases.

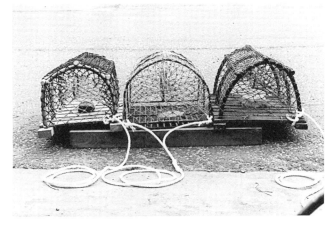

Fig 35 Left to right, wooden base with cane frame, steel frame and base plastic coated, wooden base with plastic pipe frame

79

Water piping has one disadvantage; in the standard wall thickness which is comparable in price to cane, the framework can be distorted when drawing the 'eyes' tight. Neither can plastic be preformed like cane to give the flat topped shape which is useful for easy stowage on board. Water pipe designed for high pressure with a greater wall thickness overcomes the problem of distortion, but is much more expensive than that of standard wall thickness.

Even more durable than piping is the solid 19 mm ($\frac{3}{4}$ in) diameter plastic core around which high pressure hydraulic hose is constructed. This material is occasionally available from the makers of hydraulic hose and is virtually indestructible either by the process of hauling or heavy seas.

Synthetic materials Today synthetic fibres have replaced the one time natural materials used for the netting to cover creels and pots. Without exception these new materials are superior in all senses to those once in use throughout the fishery. While nothing is abrasion proof in a hostile environment such as the sea-bed, being resistant to rot, synthetics have a much longer life than natural fibres.

Synthetic twines for trap sheeting can be of either twisted or braided construction, while many fishermen still knit their sheeting from balls of twine, there is an increasing trend to purchase a length of machine knitted netting and cut off the number of meshes required. For the standard 760 mm (30 in) by 460 mm (18 in) creel this would be 16 meshes long from a sheet 18 meshes broad with a mesh size of 70 mm ($2\frac{3}{4}$ in) stretched mesh.

Dimension of entrances Opinion varies on the best method of constructing the entrances or eyes, but on the east coast at least it is agreed that they should be of a minimum size of 5 inches in diameter. A dimension generally tested by passing the clenched fist through the eye, it should not touch either side or top. While this size is sufficient for

80

the crab stock on the east coast, in the Channel grounds it is required to be much larger, up to 7 inches.

Many fishermen prefer to make the hard eye themselves from split canes or plastic coated wire, oval rather than round shaped. These oval eyes are considered to allow larger crabs to enter, a similar advantage attributed to the soft eye which contains no rigid frame. In fact there is little to choose between the different types, with the preformed hard plastic eye hard to beat for sheer convenience. The soft or frameless eye is allegedly the most escape proof of all methods, but against this there is the danger that a creel which is too escape proof may prove equally difficult to enter.

The 'parlour' creel Despite what the layman's opinion may be, creels, pots or traps can never be one hundred per cent escape proof, such a creel would have no eye or entrance. One step to retain the catch is the provision of an inner compartment or 'parlour' in the creel. This is made by working an eye of netting on the centre bough leading into a blank chamber. The theory is that once the catch has had a taste of the bait and hunger satisfied, he decides to move on. The easiest way for it to go is through the broad entrance of the parlour rather than through the narrow end of the eye leading to the outside and freedom. Once in the parlour there are two exits to negotiate instead of one. Hopefully as it

Fig 36 Frame with parlour knitted on

81

attempts to escape from the outer compartment it will once more wander into the parlour. The holding properties of the parlour can be even further enhanced by having a flap of loose netting on top of the eye, making the parlour at least almost escape proof.

Parlours are well worth the extra effort. Most trap fishermen find that when they have been unable to haul their gear for several days, the catch is taken mainly from any parlour creels which are fishing. For part of the fleet, or at least interspaced along its length, some parlour creels can pay dividends. With the standard length of creel it is only possible to have a single eye and parlour; the snag of one eye is that it may be blocked by weed or other obstruction when it settles on the sea-bed. This can be overcome by making double eyed parlours by increasing the length by a half, or even double eyed and double-parloured twice the length of a standard creel.

In the ordinary creel without a parlour some action is

Fig 37 Double parlour creels

possible to prevent the escape of the catch. The most common method of doing this is to bend a U shaped piece of wire around the top of the eye. The two lower prongs of the U should be of sufficient length to touch the opposite end of the eye. Care should be taken when fitting these escape inhibitors to ensure that they do in fact hinge inwards to allow the catch to enter. Effectiveness of the inhibitors must be much impaired when the creel is lying at an angle allowing the 'shutter' to swing open.

When knitting in the eyes of a creel there should be sufficient meshes to reach just over half the width of the creel when the eye is pulled tight. An eye which is too short will make it easy for the catch to enter and escape, one which comes too close to the opposite side of the creel will prevent the larger specimens from entering. What is vital is that the position and depth of the eye does not allow potential quarry to reach the bait without first entering the trap.

Materials used in a standard wooden creel are extremely buoyant, and while the creels are heavy when soaked, some form of weight is needed to sink them initially. To this end weights must be lashed securely inside the base. Secure is the word – an insecure weight acts like an internal battering ram on the trap in any amount of sea. At one time it was only hard volcanic stones which were tolerated as creel weights. Materials such as concrete, brick, iron or even sandstone being viewed with suspicion by fishermen if not by shellfish. Most of these prejudices have now gone, for speed and security there is nothing to beat the preformed concrete weight, much simpler to tie down securely than uneven odd-shaped beach stones.

Retaining bait Whether the trap is of inkwell or creel shape some method of holding the bait securely within it is necessary. In the former this is mostly done today with a section of car inner tube stretched tightly across below the funnel or neck. Creels usually have the bait

83

band fixed centrally from the weight to the top of the middle bough. Braided twine doubled is the best material for bait bands and they should be tied as taut as possible. Assisting to hold the bait in position is a stopper knot, slid up for inserting the bait and jammed down tight to retain it within the creel. Instead of a stopper knot, a scrap piece of leather, plastic or rubber, pierced with two holes can be used in a tight fit on the bait band twine, this latter method often being quicker to work with cold or gloved hands.

Fig 38 Flat fish in bait band

One further item is required to complete the creel, a means of inserting bait and removing the catch. Simplest form is the lace up door. Here a line of meshes is cut straight along the longitude of the creel two or three meshes up from the base. These are doubled up with twine for strength and a piece of strong braided twine, heat sealed at the end, used to lace the edges together. Much favoured in some regions is a folding door: for this the meshes are cut just below a sidestick but on one half of the creel only. A length of cane is lashed to this and it is secured over the creel by a double piece of car inner tube and wire hook. Many makers of steel frames also offer a folding door. This is in fact the entire end of the creel which

84

can fold downwards and is secured again with inner tube and wire hook.

Constructing a wooden creel

Making a wooden creel is a labour intensive affair and any short cuts to speed production are worth while. Construction is best on an assembly line system, *ie* nail up a number of bases in one session, drill, tie in weights *etc*. A hammer saw and pocket knife are considered to be sufficient tools for making a creel, with the addition of netting needles for working in the eyes. A small gas blow lamp is a useful accessory for heat sealing all the cut ends of twine to prevent them opening up when fishing. Drilling the bulls to take the canes was at one time done with a heated iron from the domestic fire, an auger or hand brace can be used, but even a small electric drill speeds this tedious process.

Some treatment has always been given to creels, mainly directed at prolonging the life of the components. Any wood immersed in the sea is always prone to the attention of marine boring worms, and it is to the preservation of the base that most treatment is directed today. When natural fibres were used for netting they also needed protection against rot, the process giving 'life' to the netting.

Coal tar was a favourite used for many years, as can be borne out by the still tarry state of some harbours where this treatment was carried out. Bark or tannin largely superseded tar, even when local gas works were still in use and could supply that material.

Bark is supplied in block or powder form and must be mixed in water to make a solution for treatment. A pot or tank capable of immersing the creel, and provision for heat, is needed as the mixture is best applied hot and requires boiling to dissolve the bark. It is a satisfactory feeling to view a fleet of newly barked gear of uniform colour; it remains a task which most fishermen still relish.

Fig 39 Barking creels helps to preserve component materials

Products used in the building trade and similar to coal tar can be bought for creel treatment, but the price of them is becoming prohibitive. At £25 for five gallons even when diluted 2 : 1 with paraffin makes it expensive, when a similar sum can provide a hundred gallons of bark solution.

Opinions differ as to the merits of any creel treatment; those in favour say that it 'sweetens' the creels and makes them fish better. Those who do not give any treatment maintain that a week ashore and a few showers of rain has the same effect. What cannot be denied is that wooden bases have a much longer life when treated with some form of preservative.

Caution should be employed when experimenting with any new products. For instance, creosote, while it would preserve the wood, seems to be intensely disliked by most shellfish.

No matter what materials are used in the construction of traps, or the precautions taken to preserve them, the ongoing work of hauling and shooting from the first day they are in the sea begins their deterioration. Wedged between sea-bed boulders or caught on obstructions, inevitably some will be bent or broken hauling them free. That is without them being caught in shallow water with a sea wind, where years of wear can befall a trap overnight. If the fishing is not pursued daily, large numbers of crabs can wreak havoc on the netting coverings, a problem less acute with modern twines.

Stages in making a traditional wooden creel.

Fig 40 Drilling holes in 'bulls' to take boughs

Fig 41 Base nailed up, drilled, weights tied in, bait band in position. The needle has been used to thread a piece of plastic on as a 'stopper'

86

Fig 42 A simple former for preshaping cane boughs

Fig 43 Softwood wedge driven into end of cane to secure it in bull

Fig 44 Lashing and nail for attachment of bull to sidestick

Fig 45 One method of securing netting on end of creel

Fig 46 An alternative method of securing netting on end of creel

Fig 47 First cut for forming the eye

87

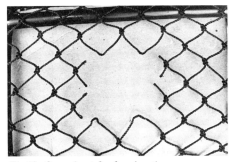

Fig 48 Second cut for forming the eye

Fig 49 Eye knitted in to reach just over half-way across the creel

Fig 50 Scrap wood used to prevent chafing of the twine securing the weight

Repairs to creels Some time, or rather several times, in the life of a creel repairs are going to be needed. Perhaps it may only be the repair of a broken mesh in the netting or it might go as far as replacing a wooden base in a polythene-pipe framed creel.

Wooden constructed creels lend themselves to repair with simple hand tools. Many of the repairs can be awkward and frustrating, for example to replace a centre cane bough on a wooden creel first requires the partial demolition of the remainder.

Relashing sidesticks to boughs when the netting is in place can be a particularly finicky task. Often it is better to cut the leg of the mesh which is preventing the twine from being freely threaded through, repairing it later when doubling. Sidesticks on pipe creels are less trouble, in the home-made variety at least it is a simple matter of replacing the nut and bolt and tightening.

Usually on traps with the netting outside'the frame unprotected by any armouring, one of the first repairs will likely be to rubbed meshes where they stretch across the frame, the corners being extremely liable in this respect. Only a mesh here and there may be completely rubbed or 'chipped', but the repair is best done to take in the complete side of the creel. This in effect, where the meshes are still whole, doubles the thickness of the twine, if one side requires this attention it is worth while continuing around the entire creel wherever the netting comes in contact with the frame.

Steel-framed creel repairs

Steel-framed creels are a different matter, while repairs to the netting can be done as in wooden traps, working on the frame itself is more complex. Any framework repairs to steel involves some kind of welding – a simple enough job in itself, but it may require the removal of any armouring and the netting before it can start. Even in the most modern forms of electric arc welding, sufficient heat is generated to ignite or at least melt the netting covering.

A great deal of the creel fisherman's time can be taken up with gear maintenance; in general, the older the creel the more frequent the attention is going to be. A careful look must be taken at any veteran trap ashore for repair. Is it worth the effort needed to bring it back into sound fishing order; is the time which is going to be spent fully justified? Sometimes it is more economical to scrap the lot and start from scratch on a new creel, time being better spent than on an 'old stager' where further work is going to be needed in a few weeks' time.

Inevitably wear and tear on creels does not take place on a uniform basis. Rarely are all the trap's components in an equal state of decay and destruction. As often as not a broken battered base will be

89

surmounted by a frame and cover in pristine condition. In such circumstances it is possible, if traps have been made to a standard size, to join the two parts together to make one fishable trap. Alternatively the base can be removed from the frame and the new part fitted as appropriate.

When fitting a brand new frame and cover to a used base, what will most likely happen is that the base will now wear out long before the frame. New bases with used frames and netting are usually a better proposition, as in normal circumstances the majority of wear is to the base.

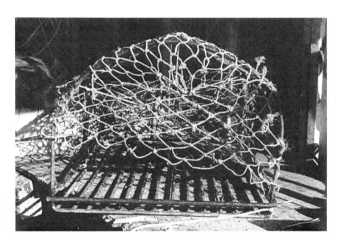

Fig 51 Even the weight of a steel creel does not make it immune from damage

90

6 Rigging creels to main rope, initial shooting and working the gear

With a collection of sound, well barked gear assembled, the next consideration is getting them into the sea, or shot, as the description goes. It is a daunting prospect to leave several thousand pounds worth of gear and countless hours work to the none too tender mercies of the sea. Single end creels (that is a creel fished on its own with individual buoy rope and marker) are seldom worked by full-time fishermen. An exception might be a large double parlour creel dropped on an isolated patch of ground. A single creel of wooden construction needs to be well soaked before shooting if it is not to float away before it becomes waterlogged.

Methods of rigging Around the country different names are given to the assembled gear shot together – fleets, strings and tiers being the more common. Joining to the main or back rope is done with strops or bridles. Strops, usually a lighter rope than the main rope, should be of a length sufficient to bring the creels to the surface as the joining knot to the back rope comes over the hauler head. This saves the effort of dragging the creel from the water, the length of strops will vary from boat to boat depending upon freeboard and hauler position.

91

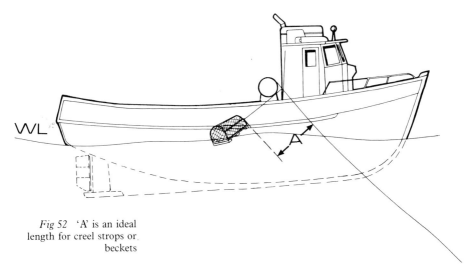

Fig 52 'A' is an ideal length for creel strops or beckets

Strops can be bent to the creels singly or doubled. For example, a boat with the hauler situated on the starboard side, a single strop bent to the creel on the same side as the door but at the opposite end, brings the creel conveniently to hand as it breaks the surface.

Bending to the creel can be either to the upright bough with a clove hitch, or a combination of eye splice looped through the end slat then clove hitched to the bough. Double strops are bent to the creels in the same way as above and the two legs joined with a splice. With double strops the chances of recovering a creel broken out of the ground are twofold in the event of one leg of the bridle being chaffed or cut.

In making up the size of the fleets to be fished, the main criterion is the carrying capacity of the boat. Before deciding upon the exact number to be fleeted it is worth considering whether to have the maximum number which the boat can hold or, by keeping the fleet smaller, make it possible when required to carry two on board. This can be a vital matter when gear has to be moved offshore into the safety of deep water with the onset of bad weather. In these circumstances the ability to boat up more than one fleet can be a great asset, saving time and effort when both are at a premium.

Main or back rope needs to be of a strength to match the weight of the boat used in hauling. For most vessels working in British waters this will mean a synthetic fibre rope of between 8 mm and 14 mm in circumference. Today the majority of artificial fibres used in the construction of rope for the fishing industry have a lower specific gravity than water, meaning that they will have tendency to float up from the bottom.

It is this buoyancy of the rope which is one of the principal causes of creels being rolled into heaps and bunches by heavy seas and strong tides. In fact, the thicker the rope the greater is the flotation and movement. When natural fibres were used in rope construction they had a specific gravity similar to water, and moving and bunching of creels was virtually unknown. Hugging the bottom and gripping the ground this rope assisted in holding creels in position rather than cause them to move.

Due to the tendency of these natural fibres to hold the bottom, they were also subject to abrasion and scouring on the sea-bed, being frequently cut on rough ground. Synthetic fibres, while they can still snag, float up in bights between the creels, keeping clear of the ground. Available today are ropes of man-made fibres specifically designed for pot and creel fishing, and like seine net rope, they are weighted with a lead wire interlayed into the rope. This rope will hold the ground in the manner of natural fibres but have a much higher resistance to abrasion. On the severest kind of ground they are prone to fasteners and their ground holding properties are best appreciated on the more open, less rugged, sea-bed.

Rope which is going to be hauled by a self-hauling vee wheel should be no harder than medium lay with a soft hairy surface to obtain maximum grip. Hard layed

93

shiny rope such as nylon or courelene does not jam in the vee, tending to jump from the wheel with every roll of the boat; an extremely frustrating situation, breaking the working rhythm of the crew.

As in any other purchase, when buying rope the buyer gets only what he pays for. Good quality rope constructed from round filaments of medium lay will outlast soft rope composed of flat filaments. While the latter is sufficient for dahn lines, constant rubbing across hard ground soon wears and abraids the fibres. It is purely a matter of choice. Top quality rope can last up to six years' of full-time work, the cheaper variety maybe two or three. Costing probably works out about even but the risk of losing gear through breaking weak rope should always be borne in mind.

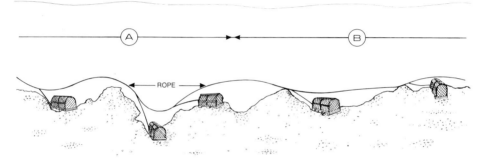

Fig 53 Creels shot slack as shown at 'A' will allow buoyant synthetic rope to float upwards reducing the risk of cutting or fraying during bad weather. Shot tight as shown at 'B' the risk of broken ropes is greater

Bending strops to main rope Having decided on the number of creels and selected suitable rope, the next step is to bend the strops to the main rope. Several methods can be used in doing this, very much dependent on individual whims. Where self-haulers are used, the fewer knots and splices on the main rope the smoother the hauling operation, with no reason to stop when lifting knots or splices through the wheel. Creels bent to the main rope with an

94

overhand or thumb knot give the smoothest hauling rhythm.

Hazards of proximity to other boats A main rope with no knots or joins has disadvantages when working in close proximity to other boats on the same grounds. Sooner or later, through fog, accident or other causes, the situation will arise when two boats' gear is shot across or one above the other. When this happens, if there are no knots to untie to pass free end over the offending rope it becomes necessary for someone's rope to be cut, a course of action upon which most fishermen are loathe to embark upon.

Because this sometimes happens, some fishermen make up their main rope in 10 fathom lengths with an eye splice at either end. These are bent together leaving one eye protruding for the creel strop. If other boat's gear is snagged it is then possible to haul to a knot, unbend creel and eye splice of own rope before passing free end over to clear away. Likewise it is possible to make up the main rope tying the creels on with the thumb knot, but having at intervals reef knots with the free ends tucked back into the lay. In theory these can be unbent to clear, in practice after a few weeks of hauling, these knots are so tight that the knot must be cut and retied.

Without the use of weights or anchors, wooden creels at least would have the inclination to move across the sea-bed. On the main rope it is usual practice to leave an extra two to four fathoms between the last creel and anchor. Proper anchors can be used, but on the hardest ground these are likely to snag with consequent loss. A suitable substitute can be had in the shape of railway chairs, or on the severest ground, a bunch of scrap chain. Whatever is employed it should be rigged with heavy hard laid rope with an eye splice or loop for bending to the main rope.

The use of dahn buoys to mark static gear To enable its recovery and inform others of its location all forms of static gear must be marked with flagged dahn buoys which must be marked with the boat's name or registration number. By far the easiest colour to pick out under most light conditions is a flag made from the darkest coloured material possible, be it cloth or plastic. Black can be seen at a greater distance than any other colour whether the background is land, sea or sky. Colours like bright yellow and even fluorescent orange can in some light conditions prove difficult to see at a distance. Hard to beat is the jet black flag set on a staff keeping it at least 2 metres (6 feet) above sea level. Buoyancy at the centre point and weight at the base are needed to keep the flag staff afloat and upright in the water.

Additional flotation to that on the flag staff is needed to keep it afloat, this being best attained with a 20 in inflatable plastic buoy, or a hard plastic trawl float one to one and a half fathoms from the flag.

All that remains now is to join the flag to the bottom

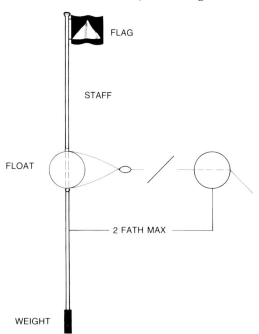

Fig 54 Marker dahn buoy

gear at the anchor or weight. A lighter rope than the main rope is used for the buoy line, thinner rope offering less resistance to strong tidal flows pulling down the marker flag. It must, however, be strong enough to withstand the weight of the boat in the event of the anchor or some of the first creels being fast to the ground.

The length of buoy rope

Where strong tidal flows are encountered, the length of the buoy rope might need to be up to three times the depth of the water being fished. Where coastal traffic is heavy the dahn and rope are liable to being run down and lost: most likely to happen with the changing tide when the rope swings slack upon the surface. A fine balance must be struck between enough rope to support the dahn while at the same time not being prone to loss.

Dry wooden creel need soaking time

Despite being weighted, dry wooden creels will float when first shot. Initial shooting needs to take place well away from any chance of them being cut by coastal traffic. Shooting at right-angles to the tidal flow with the main rope pulled as tight as possible, plus extra weight added in the centre of the fleet will sink the creels in two tidal cycles. This initial shooting across the tide brings the weight of the flow against the rope, forcing it and the creels down much sooner than if they had been shot with the tidal flow.

New wooden creels with cane framework can take up to a fortnight before they start to 'fish' properly. Due to their porous internal nature, canes will fizz and bubble as water displaces the air held inside. Deep water, with the extra pressure forcing water in will cut the initial soaking time, twenty fathoms being regarded as the minimum depth for the first week that new wooden creels are in the sea. No such problem exists in the case of steel or steel based creels, these will sink and fish immediately.

When first shot new wooden creels are best hauled

97

within two days of shooting in case any are lying with the covering netting chaffing on the ground.

Characteristics of traps The fishing characteristics of any shellfish trap depends very much upon the nature of the materials used in its construction. The presence of light swell or ground motion in the sea is a great stimulation to encourage shellfish to become active and willing to feed. Growing swell, however, does not usually provide very good fishing; rather it is when it has reached its heaviest and begins to fall away that the best catches can be expected. There will be an optimum depth at which the effect of the swell is creating the stimulation without actually moving gear across the sea-bed. This optimum depth may be only a few fathoms either side of a given datum. For example, creels in say eight fathoms may have been moving and the catch insignificant. At ten fathoms the fishing may have been good, but from twelve fathoms into deeper water the influence of the swell will have had less effect, resulting in a smaller catch than from the ten fathoms.

Considerations when A knife-edge balance exists between the reckless risk of
shooting gear gear shot in shallow water and the calculated gamble with the weather. Sometimes it is difficult to define, this is mostly so when forecasts give seaborne weather yet the winds come from the land. At peak fishing times some catch will still be made in the deeper water, where the creels are placed for the forecast weather. Much better results will be had from creels placed in a depth of water to suit the actual weather. Larger numbers of shellfish are taken and large grossings made by fishermen who take these risks. On the debit side large numbers of creels are lost each year by the same fisherman. When the replacement of gear is taken into account at the end of the year the man who has been more cautious with his gear will be as far ahead as the gambler who has lost.

When trap fishing out of season, it can be a totally

futile task hauling traps which are in the wrong depth or over the wrong ground. In some winter fishings if risks are not taken the gear is just as well brought ashore for overhaul.

Heavy steel creels of 10 mm ($\frac{3}{8}$ in) round bar will fish in considerably shallower water than wooden ones without moving and disturbing the catch. Steel creels braided with cut tyres will be more liable to movement than plain steel, while combined steel and plastic will be in between. These are all considerations to be taken when deciding which depth is to be fished.

Crew duties When it comes to shooting the gear, crew duties are more clearly divided. One man will control the boat, adjusting speed to suit the operation and steering the boat so that creels fall on the most productive ground. In a two-man crew the second man will shoot dahn and anchor, followed by the creels, back over-board in the reverse order to that which they were hauled in. If weather conditions are poor this is not always a simple task of selecting the next in line creel to go over the side. With the vessel rolling the stowed creels may have tumbled over, upsetting their order on the main rope. A creel shot in the wrong order means a foul shoot, with the rope drawing more creels overboard resulting in chaos.

In larger than two-man crews, one will control the boat with the other standing at the shooting position, with other crew members handing the creels to him. A single-handed operation means that there must be some means of controlling the boat from the shooting position. Not only steering but a means of stopping and going astern should be at hand for safety in the event of a limb becoming entangled in the rope. Single-handers are always advised to have a strong sharp knife ready to cut themselves free in real emergency situations.

Fig 55 Shooting creels on a boat with forward stowage

Reshooting the gear must take place over the same side of the boat where it was hauled. While it would not be impossible to shoot on the opposite side, it could be inconvenient perhaps with the need to move spare bait and catch boxes *etc.*

Avoiding shooting on lee side In strong winds the boat must be set up to ensure that shooting is not going to take place on the lee side of the boat. At the slow speeds which shooting takes place a considerable amount of leeway is made, driving the boat over the top of rope and gear. A likely consequence of this would be that the rope, even when fishing fast-sinking steel creels, would be drawn into the propeller as the boat 'crabs' over them. Avoiding this would be possible by shooting over the stern, but this is a method little used in trap fishing.

WIND
DIRECTION

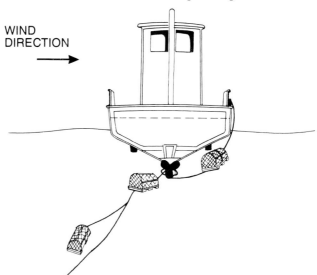

Fig 56 Shooting creels on the lee side can result in rope and creels being drawn into the propeller

While in most instances the rope should be drawn reasonably tight when shooting, where high pinnacles rise from the sea-bed this would have the creels suspended between them like washing on a line – a totally futile exercise when trying to catch bottom-dwelling shellfish. One way of ensuring that the gear lands slack is to steam the boat against the tide as the

Fig 57 In extreme cases creels shot on a tight rope may be held clear of the sea-bed operation is taking place, but maintaining a speed over the ground to ensure that the rope and traps are not falling in a heap on the bottom.

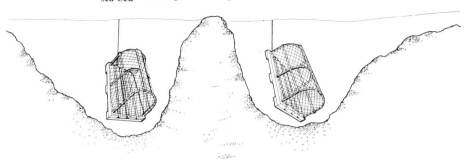

Ensuring creels fall on suitable ground Again to achieve a slack shoot the creels can be bent or faked back and forth across the ground – also a handy method which drops the creels into likely lobster-holding crannies – a tactic most suitable for placing a full fleet on a small narrow piece of ground, which if they were shot straight would overreach the edges.

Fig 58 Shooting creels at right angles to a strong tide as shown by boat 'A' can result in creels being carried away from suitable ground

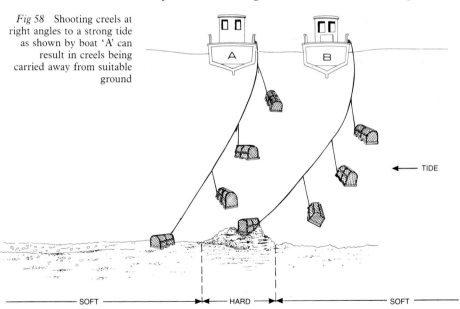

Where only small productive patches are found on otherwise unsuitable ground, every effort needs to be made to get as many creels as possible to fall upon it. On this ground by shooting at right-angles to the tide, the helmsman watches for the patches of hard showing up on the echo sounder. As these patches appear, the boat is steered up through the tide across the hard allowing as many creels as possible to fall on it.

Isolated patches of ground can be covered by starting in the centre and shooting in ever-increasing circles: a method best used for slack tide, as the creels would fall down-tide one upon the other in any amount of tidal flow. Even smaller boulder heaps and large stones can be covered by dumping several creels over them without waiting for the main rope to draw tight. These small patches should not be ignored, while part of the fleet may be wasted on soft ground, the size of lobsters will often more than compensate.

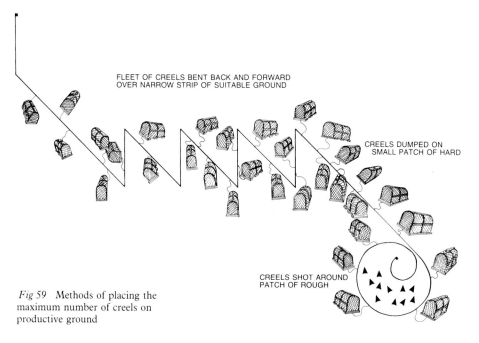

FLEET OF CREELS BENT BACK AND FORWARD OVER NARROW STRIP OF SUITABLE GROUND

CREELS DUMPED ON SMALL PATCH OF HARD

CREELS SHOT AROUND PATCH OF ROUGH

Fig 59 Methods of placing the maximum number of creels on productive ground

102

Gear which is shot slack, bent back and forth or dumped on isolated patches is the more liable to bunching up in heavy seas and strong tides. Tactics such as these are best reserved for neap tides and used only when weather conditions and forecasts are favourable.

Creels shot at right-angles to the tide brings the added problem in that they will land a long way downtide from the shooting position, perhaps missing the intended target altogether. The stronger the tide and deeper the water the greater the distance of drop. Due allowance needs to be made for this when the boat is positioned at the commencement of the shoot.

Hauling technique   Hauling a fleet of gear commences with picking up the marker flag and engaging the buoy rope in the hauler. If tide and wind are strong the boat will need to be steamed ahead until the anchor is brought on board. Assuming that there are two crew members, one will operate the hauler, steam the boat ahead and lift the creels on board. If a self hauler is being used he will have plenty of time to open doors, remove some of the catch and clean old bait and rubbish species from the creel. During heavy fishing of crabs he may only have time to remove part of the catch when several large crabs are jammed in a corner. At other times when lobstering with no rubbish species in the creel, it is possible to remove the catch and slip a piece of fresh bait in through the eye, eliminating the need to untie the door.

In normal fishing the second crewman will be involved in baiting, fastening the door and stowing the creel. Heavy crab fishing would see him removing part of the catch, baiting, fastening the door before stowing. Light fishing would mean that his job would be removing the catch and baiting every other creel before stowing.

Where larger crews are employed, the work is further divided even to the extent of having one man

103

in the wheelhouse steaming the vessel ahead when necessary. Crewing single handed means that the operator is going to have a busy time performing all the functions in the above paragraph on his own.

If the boat is fitted with a capstan head a further operation is required when hauling the gear. As before, the sequence commences with the marker dahn being taken on board and the hauler mechanism engaged. In this instance, instead of the rope being placed in the vee wheel, from one to three turns are taken round the capstan head. With a capstan head a crew member must pull the rope away from the capstan as the rotation of the head brings it on board. Gathering the rope in loose coils, it is flung down behind him in a position which is convenient for shooting. As the strops come to the hauler the operator must take the turns from the capstan, lift the creel on board and re-engage the turns before hauling continues.

Fig 60 Hauling with capstan head hauler positioned forward of the wheelhouse

Bow and stern hauling While most boats haul from the bow, pulling the boat ahead all the time, many of the cobles on the northeast coast of England still haul by the stern. For this the rudder is first unshipped, leaving the hauling sequence to be carried out as in a conventional vessel, but stern first.

Whatever method of hauling or type of hauler is used, the work is done at the fastest speed possible. Several factors come into this, one being that there are a set number of creels to haul in the day; by working at maximum speed this will make a shorter working day or allow time to prosecute some other method of fishing. By hauling at the highest possible speed the boat will keep ahead motion or 'way' upon her as it is hauled along the fleet. In keeping the boat going ahead without resorting to steaming, a smoother operation is maintained with the gear coming on board at more or less precise intervals. This allows the crew to work in unison and leads to a smoothness of rhythm which cannot be had when one crew member has to break from trap servicing to steam the boat.

What cannot be ignored, either, are the other boats engaged in the fishery – work at a leisurely rate will make it certain that there is less choice of ground to shoot than by those boats which have hauled as quickly as possible.

Direction of hauling There are two main factors which influence the direction which creels are going to be hauled: wind and tide strength, and direction. Ideal circumstances for most boats would be to haul against the run of the tide on the weather side of the boat. Without the use of 'spinners' creels which are perpetually hauled with the run of the tide will result in the rope on the strops spinning up into a tight bunch. Working on the sheltered or lee side has the boat blowing across the gear, resulting in creels being damaged by the main or bilge keels.

Ideal circumstances are seldom found, and most

105

Fig 61 Plastic spinner is essential if fleets of creels are always hauled in the same direction as the tide. They also help prevent loss of creel due to strop unlaying and chafing during bad weather

boats will lie to their gear in an individual manner. For example, hauling on the weather side with an aft wheelhouse results in the boat being blown off by the stern: with a forward wheelhouse it would be by the bow. When the wind is having a greater influence on the boat than the tide it will be necessary to haul with the tidal flow.

Only experience reveals which set of wind and tide conditions suits the direction in which the gear is hauled for any one vessel and its deck layout. Sometimes no matter which direction is chosen, creels will be coming awkwardly to hand from under the boat. At other times in slack tides and flat calms hauling can be from either end of the fleet without causing any inconvenience.

Fig 62 One method of rigging spinner; creel is left free to turn against stopper knot to the left of spinner

Frequency of hauling baited traps

Hauling baited traps takes place at intervals very much dependent upon the season of the year and the current catch rate. In general at the height of the season when fishing is good, hauling every twenty-four hours will result in the heaviest catch at the end of the week. Seldom when catches are good does leaving the traps set for two nights result in double the catch than for

106

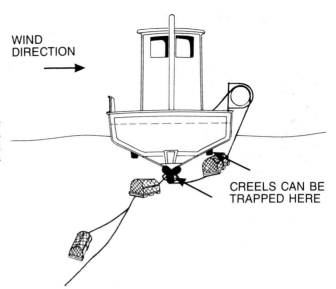

WIND
DIRECTION
→

CREELS CAN BE
TRAPPED HERE

Fig 63 Hauling on the lee
side can result in creels
being broken when trapped
by the main or bilge keels

one night. This is more so when the gear is being worked in shallow water during summer than when in deeper water offshore in winter.

Cold water in winter and a poorer catch rate often make the twenty-four hour cycle of hauling unprofitable. With the shellfish's metabolism slowed down by the cold it can take longer for the tide-borne scent from the baited trap to encourage them to enter. In winter and while fishing in deep water, the forty-eight hour cycle can lead to an increase in the weight on the tally sheet come settling day.

Creels to be cleared of old bait Old, soft and rotting bait must always be removed from the creel; any pieces floating free inside distract any potential catch from entering through the eye. Any large pieces washed against the outside netting will allow shellfish to feed from the outside rather than enter the trap. For similar reasons the new bait must be held firmly by the bait band and placed in such a position that the creel must be entered before it can be grasped.

107

With a full fleet on board and serviced, broken creels repaired or replaced, the boat is steamed to new ground ready for shooting to begin. If the fishing has been good, this might only mean turning the boat and allowing the change of tide to drop the creels on new ground. Should the catch have been poor, or there is a change in sea conditions, considerable steaming may be necessary.

Dealing with fasteners Inevitably when hauling creels gear or rope will become held fast on the bottom. Fasteners are presumed to be either creels jammed into rock crevices or fouled by wreckage. It may in fact not be a creel which is held but the rope pulled or looped under boulders or rocky protuberance. An initial step to clear a hold is to steam the boat up through the tide while shooting away several creels before making the main rope fast to the boat. Towing uptide after this gives a paravane effect to the trailing gear, while the extra elasticity of the rope helps to reduce damage. Should this fail, a short up and down hold while surging the rope, hauling tight then letting it spring back may clear it. Sometimes, surprisingly, with the rope removed from the power of the hauler and jerked by hand it comes free when all else has failed.

Holds which persist need the gear to be reshot, in the direction from which it was hauled, and an attempt made from the other end. In most instances this opposing pull will result in clearance. In the extreme case, other than reshooting and trying on a different tide, nothing remains but to make the rope fast to a strong point on the boat and tow until the creel breaks or the rope parts.

Great care needs to be exercised in some conditions when breaking fasteners from the ground. Little risk exists in fine weather and slack tides, even when working on short holds between boat and obstruction. With a powerful hauler and light boat it could be possible, even in ideal weather conditions, to pull the boat rail under. Attempts to free persistent fasteners when any swell is running should be made only with several creels away from the boat.

At one time a different mode of working was used in winter for crab fishing on the east coast. 'Over running' did not depend upon precise placement of the creels but relied on the daily difference in the tides to drop the gear on fresh ground.

When over running, a larger crew (a minimum of three) was needed and creels were worked in fleets of up to a hundred. This method consisted of taking dahn, anchor and the first few creels on board, serviced and baited. As hauling progressed one man would shoot way the creels in sequence as the rope drew tight with the boat hauling ahead. The merit of this method was that with the gear hauled it was also shot away, making a considerable saving in time. Over running is not much used today; many of the grounds where it was pursued are now fished by mobile gear methods and today's smaller stocks need a more exact placement of the gear for successful fishing.

Fig 64 Creel held fast by rock at 'A' can best be freed by steaming into the tide. 'B' shows how buoyant artificial fibre rope floats upwards between creels

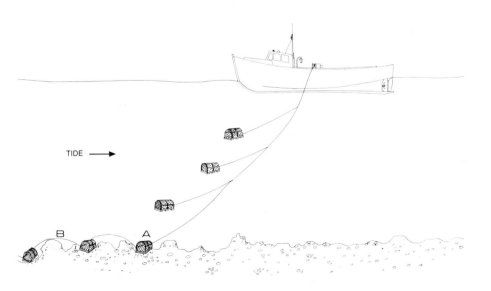

TIDE ⟶

B   A

7 Hazards, weather and the work of other boats

There can be few other jobs or professions where the weather can have such an influence on the success or otherwise of the operation as the commercial fishing industry.

This is especially so in the case of those who fish static gear such as in the crab and lobster fishery. Not only must the current weather be taken into consideration but also that which is about to transpire within the next twenty-four hours. Nor does it merely concern the local weather; on exposed coasts conditions many miles distant will have a considerable influence on the state of the sea in inshore waters. While calm settled conditions may exist overhead, gales blowing coastwise on distant sea areas can have breakers rolling across harbour mouths, keeping boats pinned in harbours, perhaps with vulnerable gear at the sea's mercy in shallow water.

This need to fish in shallow water, where some of the best catches can be made, makes the creel fisherman's job one of the most hazardous of the sea. There will always be days when larger boats are tied up snug in harbour, when the static-gear man must if possible go to sea in an attempt to get his gear into safe depths of water. Sadly, more vessels are lost by contact

110

with the land than by the sea; working in close proximity to rocky shorelines doubles the risk in this respect for the creel fisherman. Even in fine settled weather the most experienced of fishermen can suffer the indignity of hauling his boat ashore, to be left high and dry with an ebbing tide. If hauling could always take place on a filling sea, this hazard would be eliminated. With the world being far from a perfect place, inshore hauling must be done at times when time and tide wait for no man.

Even the most strongly constructed of creels is but a fragile thing when exposed to the full fury of the sea. With this in mind the fisherman must seek sufficient depth of water to safeguard gear when onshore weather threatens. In a prolonged spell of gale force sea winds, twenty fathoms depth may be only just enough for wooden creels. With a north or northeast wind on the east coast, each tide will find seas building up, breaking upon themselves and surging right to the sea-bed. At times like this fishermen stand at the pier head watching the destruction being wrought and wonder if it is all worth while. It can be severe enough in summer, but the cold heavy water of winter falling right to the bottom carries all before it, blasting all in its path with a shower of gravel and stones.

Weather forecasting services An excellent weather forecasting service covers all the waters in the UK. Regular broadcasts are made by national and local radio and television stations plus 'Marineline' and 'Weatherline', which are only a telephone call away, with numbers in all local telephone directories.

Yearly weather patterns and local weather lore should not be ignored; for instance, in southeast Scotland the 'Coronation breeze' is still spoken of with some awe. Likewise 'the Gowk' April Fool, the first week of April, generally sees a spell of bad weather. Good weather seems to follow similar trends with the week prior to Christmas, usually being one of fine weather with only light west winds.

Choice of ground in bad weather When moving offshore into deep water the fisherman has the choice of shooting over hard or soft ground when bad weather is imminent. On the hardest ground the effects of heavy weather will wreak the worst damage, rubbing and chaffing creels against boulder and reef. Softer smoother ground will fail to hold the gear against the surges and can lead to the entire fleet rolling into one tangled mass. As on the hard ground, equal damage can be caused by this, not forgetting the arm-stretching, back-breaking task of getting the bunched gear on board and untangled.

Where possible a mixed type of sea-bed should be sought with an effort made to drop the anchors or weights on good holding ground. Extra weights added to the anchors also help, as it is from the buoy lines that movement and consequent bunching usually take place. As the rollers lift the dahn, often pulling at the anchors or stones, the tide assists each surge, pushing creels down one on top of the other until all are in one

Fig 65 Heavy weather can create havoc with any form of static gear, including pots or creels

112

heap. When the direction of a blow and consequent seas is forecast, gear shot at right-angles to it will have less chance of rolling together than one shot in the same direction. Shot across the sea, any movement will be in an arc, hopefully carrying the creels away clear of each other.

While moderate amounts of swell will stimulate shellfish to feed and therefore enter a baited trap, heavy seas disturbing the bottom sends them post haste to the deepest securest rock cavity they can find. Not all make it, and when lobsters are close inshore, heavy seas can wash many ashore dead. Extremely heavy seas at the midpart of the lobster season can drive them from the inshore grounds until the following year.

Effect of river water entering sea

Where large rivers enter the sea, times of heavy rainfall bring a mass of dirty cold water, and the catching of shellfish can be severely affected. Bad enough when the fresh-water comes direct from rain, but snow melt is ten times worse. Crabs seem to be more susceptible than lobsters to this sudden drop in sea temperature and will apparently vanish from the grounds. Doubtlessly they will still be there but their metabolism is so slowed by the cold that they will be dug into the sea-bed, emerging only when warm weather sharpens their appetite.

The risk of static gear by mobile gear

When creels are moved offshore seeking the sanctuary of deep water, another matter other than the weather must be taken into consideration. On grounds where mobile as well as static gear is worked at the same time, the chance is always present of losing the latter by the movement of the former. Any gear which is towed mixes with shellfish traps as oil does with water – not at all. At one time there were plenty of offshore rough patches in deep water where gear would be safe in bad weather. Technology in the fishing industry has advanced since those days, in electronics and navigational aids as well as the rigging of nets. These

113

techniques have opened up grounds to trawl methods which would have been impossible even a decade ago. Now trawlers, either single boat or pair team, can fish safely on grounds which were previously the preserve of the shellfish operators.

Retrieving lost gear

Sooner or later, either due to bad weather or by being fouled by coastal traffic, both dahns on a fleet of creels will be missing. It is for this reason that it is vital to always have a sharp shore mark or Decca reading of the position of the gear. On open ground a grapple can be towed to recover rope and gear. Prime lobster ground, sharp and peaked, can inhibit the travel of a grapple which will become snagged every few minutes. This is where the traditional wooden creel with the bull protruding past the slats is a useful tool. By shooting a fleet of these creels at right angles to the lost fleet the protruding 'lugs' can, on most occasions, hook up the lost rope. By hauling at slack water when the rope is likely to be floating up in bights between the creels, recovery is all but guaranteed. As the weight of the lost gear is felt to come on the rope, the boat should not be steamed ahead, allowing the hauler to keep a constant strain on the rope to prevent the recovered gear from jumping clear.

Fig 66 How the protruding lug on a traditional creel can pick up rope on 'lost' gear

When the lost gear comes to the surface a spare dahn and rope is tied to it and dropped away. Once the fleet on board has been hauled and shot well clear, the dahn is then recovered. As likely as not, the dahn will be at the centre of the fleet. In case the rope is parted the spare dahn is shot away and the remainder of the creels overrun until the end of the fleet is reached. From here the operation can proceed as normal, hauling and reshooting the gear once new dahns have been placed at either end.

Stinging jellyfish

No discussion of the hazards of any kind of fishing would be complete without the mention of one of the most painful creatures of the sea; stinging jellyfish or

114

'swithers' as they are called in some parts. Blue jellies are comparatively harmless; it is the red sting trailing variety which causes the most discomfort. At the best, as the wind blows them from the rope into face and eyes they are a discomfort; at worst to those allergic, when a severe amount is blown into the face it can necessitate medical treatment. A few pounds spent on industrial face masks, even if they are slightly claustrophobic, is well worth the cost for comfort.

Caution imposed by part-time fishermen

With an ever increasing amount of leisure time available to the shore-based worker, the grounds close to harbours can be intensely fished by other than full-time fishermen. Although they may be operating comparatively few creels per boat, when several are working, the grounds can rapidly become unavailable to full-time fishermen. This mainly arises because while *bona fide* fishermen may be operating in early morning, part-timers will usually work in the evening.

Most full-time fishermen have a deep respect for the gear of their fellows, going to considerable trouble to avoid cutting rope or damaging traps. Many part-timers are ignorant of or unwilling to follow this code and do not have the knowledge or experience to read the dahns and gear placements. In the morning when the professional fisherman hauls his creels he may find that every few fathoms he is bringing the part-timer's gear to the gunwale. Once clear, should he reshoot on the same ground in his previous berth, he might find that when he next hauls his rope is cut to ribbons and most likely not even retied. He is left with little choice but to move, even if fishing has been productive, rather than risk further damage to his gear. Despite successive governments being able to fetch in a multitude of inshore legislation regarding salmon fishing, none have been able, or seen fit, to take any steps to protect the precarious living of the inshore shellfisherman.

115

8 Bait; different according to season

Without doubt, both crabs and lobsters will be enticed
to enter a creel or pot by fresh bait before any other.
For crab fishing the fresher the better – fresh will
always outfish stale or salt bait without exception.
Although many lobsters are taken in creels baited with
far from fresh offerings, it is the other factors rather
than the preference to near decomposing flesh which is
the reason.

Bait for trap fishing usually takes the form of whole
or parts of white fish. Due to the tenacity with which
the quarry attacks the bait, the firmer and tougher the
bait, the greater the catch per trap. As long as some
bait remains held in the bait band it will continue to
attract more quarry. While the oily flesh is a great
attraction, soft baits such as herring and sprats are
poor in this respect and are quickly shredded and eaten
by the first crab or lobster in the trap. Baits such as
flatfish and gurnards with firm, tough flesh have the
best resistance. Provided the bait band holds the bait
securely these will continue to attract long after other
baits have been demolished. In between are the more
commonly used roundfish, coalfish, whiting and
codling. Filleted frames from 'rounders' even if the

majority of flesh is removed, retain the attraction of the oily livers sending an aroma downtide.

Due to their shape, flat fish can be difficult to retain within the bait band once they have been attacked by the first quarry which enters. If a cut is made into the fish on either side of the gills, slotting the twine of the bait band into these cuts holds it secure until the majority of the carcase has been demolished.

Fig 67 Flat fish are best split to secure in bait band

When larger coalfish *etc* are used for bait they can be cut in half and a nick made in the back, again for retaining within the creel. Gurnards with their firm flesh, while excellent bait, are difficult to handle due to the many spines on the fish. When using them, a poisoned hand or finger is always a possibility, especially when hauling old bait from the creel.

Bait can be caught by fishermen engaged in the trap fishery, in which case it is likely to be mackerel or coalfish, which are found in the densest shoals. Usually the method used to catch bait is the string of mackerel flies or feathers, worked from a handline. Most small harbours have somewhere nearby where in season these bait fish congregate in shoals of sufficient density to make catching them by handline worth while.

In the northwest of Scotland and the Western Isles it

117

is the practice of lobster fishermen to buy or catch large quantities of mackerel during the autumn, salting them down to serve until the following year. As this is an area remote from processing facilities, where the crab fishery is of little importance, there is little handicap in using this salted bait for much of the year.

A practice once followed by larger boats was to trawl or seine net for bait before starting their trap hauling. With the increasing price of white fish this has declined, it being uneconomical to use prime species for bait.

Other sources of bait are from boats not engaged in the creel fishery, or the left overs from fish processing operations.

Bait need not be confined to fish, and in times of shortage condemned chicken and even pierced cans of catfood have been used with varying degrees of success.

While, as stated, the crab prefers bait which is twitching fresh, circumstances dictate that for lobsters this cannot always be productive. In cold winter waters with little activity taking place in any species of shellfish, it is fresh bait which takes the most lobsters. Unfortunately, during the peak of the lobster season in late summer and autumn, many ravenously hungry rubbish species and small soft edible crabs are equally attracted to a meal of fresh bait.

Salted bait   It is for this reason that at least for part of the year salted bait is essential if the bait band is not to be stripped bare before any lobsters can enter. In warm weather when the bait fish have been feeding on sandeels and herring fry, even salted bait will not keep unless it is gutted. There is a distinction between good salted bait and that which is sour and near rotting. Bait which has been properly salted will retain its colour; your nose tells you when this stage has been passed, and it becomes tinged with a vile yellow colour.

Two methods are used for salting. Whichever is

chosen, if the bait is whole or gutted it must be well washed in fresh-water and allowed to drain before salting. For the 'dry' method the bait is tipped into clean dry fish boxes and well strewn and turned with salt. This is the shorter-term method of preservation, and in warm weather it must be repeated daily. This makes a very hard durable bait, tough as leather and very easily handled with the salt removing the fish's external slime.

For longer-term storage the brine method is the least troublesome. Again the bait is washed and instead of being salted in boxes the bait is stored in airtight drums or barrels. If these can be obtained with air-tight sealed lids such as plastic herring barrels so much the better. In brine salting the bait is layered with salt and packed as tightly into the barrel as possible. After a day the salt will have drawn the body fluids from the fish and formed its own brine within the barrel. Gutted, well-washed bait will keep for up to a month or even longer in a strong brine solution provided the drums remain sealed.

Both methods have their merits and adherents much depending upon the original condition of the bait and the air temperature, and both can have some effect on the time which salted bait will 'keep'.

The use of shell bait In some fisheries soft crabs were once broken up by fishermen to inhibit others of their kind from entering the creel when aiming for lobsters. Besides being illegal, this was wasteful of the ever-decreasing crab stocks, and the practice has much declined in recent years. It is undeniable that shell has a great attraction to the lobster when he is wanting to regain the armour-hard quality of the shell he has just cast. Equally true is the fact that less 'rubbish' is likely to enter a creel with shell bait.

Crab bodies, when the meat has been picked from them, make a bait equally suitable for enticing lobsters and keeping unwanted catch out. It is the heart-shaped body part with the legs removed which is the best, the carapace or shell being less useful. As the heart-shaped body has been boiled, the remaining meat is firm and resistant to leaching by the water. This meat is tucked away out of reach of the probing pincers of green shore crabs and other unwanted guests, making some scent and body juices available for at least a daily fishing cycle. Many shellfish processors have a problem in disposing of crab waste and are glad to see at least part of it removed free of charge. Surely this makes crab waste a better proposition to lobster fishermen than the destruction of soft edible crabs for bait.

Fig 68   Crab waste make good lobster bait at some times of the year

Discouraging unwanted shellfish from entering trap

In an effort to keep rubbish species of shellfish out of the traps, some fishermen will leave undersized lobsters in the creels as 'guardians'. A trap containing a lobster seldom has any unwanted catch and it is doubtful if a small lobster will prevent the entry of one which is of legal landing size.

Some fishermen try the dodge of inserting pieces of white china or mirror in their concrete trap weights, hoping that the shiny white object will attract shellfish into the trap. A more positive approach is to make sure that when baited with round fish they are belly

120

uppermost, as with flats, which must have the white side showing.

No matter which bait is used it must be placed in the bait band in such a way that it is impossible for shellfish to reach it from outside the trap. If a claw or pincer can grasp the bait from the outside, the incentive to enter is greatly removed.

Experiments have been carried out in the past with a variety of compounded baits made up from a mixture of fish products. As yet these have not proved as efficient as the traditional baits: or rather they are more expensive than traditional baits. Further research is needed to bring them to a greater efficiency at lower cost.

9 Running expenses

Like any other business, running a boat in the shellfish industry is not all profit. Expenses occur from the daily expenditure on fuel to the annual premium for insuring the vessel and crew.

Marine and other insurance Taking an all risks marine insurance policy to cover the cost of the vessel is a prudent move, in fact it is insisted upon by the SFIA where any of their funds are involved. Likewise, if money is borrowed from a bank or other financial institution, they will expect the investment to be fully covered, plus insurance on the borrower's life for at least the sum on loan. Marine insurance is far from cheap and costs on average around £20 per £1,000 of insured value, a sum which can be reduced in most cases by no claim bonus and voluntary excess payments. What must be paid by all fishermen are contributions to the National Insurance fund, a legal obligation.

Other expenses If money has been borrowed, the sad fact is that until the loan is paid off, it will continue to incur interest. An unseen expense, but it must be taken into account when the books are balanced at the end of the financial year.

Every week, depending upon the hours worked and

the distance steamed, fuel is going to be an unavoidable expense. There is not only fuel but also lubricating oil, to prolong life and reliability of an engine: this and the filters must be changed at the manufacturer's stipulated intervals.

Where navigation and other electronic equipment is on hire, the rental must be paid whether the boat is working or not. Bad weather, holidays or annual fit out, rentals on hire equipment must be paid regardless.

Vessels of all kinds need a safe refuge to lie when not at sea, plus somewhere to land the catch and take on board fuel. Like boats, harbours also have an ongoing expense, be it in staff salaries or dredging and maintenance, and to finance these outlays charges are made on the craft which use the facilities. Where fishing boats are concerned this mostly takes the form of an annual charge for a berth plus a percentage on the value of the catch landed. This is perhaps one of the fairest methods, as in a good year, while landing charges may be high, a poor season sees the opposite effect. Other harbours charge a set fee on landing percentage and some base their harbour dues on the number of creels fished per boat.

Where harbours are run by fishermen for fishermen, those who are engaged in the industry as can be expected, will get fair treatment. When fishing is conducted from harbours administered by boards whose members are drawn from outside fishing and who know little about the sea and the needs of fishermen then a raw deal may be meted out. Here, where fishermen are in the minority, they can find that they are squeezed for the last possible penny in landing dues and as often as not are denied proper representation on the administering body.

Vehicles At some time or another most people working at the trap fishery will have a need for a vehicle of some description for shore-based work. If fishing is carried out from a port with no market for his catch or no

123

supply of bait, it is a definite necessity. Any van or car used to transport these commodities soon acquires its own pungent aroma with the slightest hint of warm weather. Open the windows, spray with whatever takes your fancy, it is to no avail, the distinctive perfume will follow the car to the scrap yard. If a socially acceptable means of transport is to be retained there are two courses open. The first is to have a separate van or pick-up for transporting the catch and bait – this involves the immediate expense of purchase, road tax and insurance. What can also be expected is a rapid deterioration in the vehicle, with the suspension strained by heavy loads and rust attacking from both inside and out. A second option is to purchase a trailer which has no ongoing expenses and does not suffer the same deterioration as a van platform.

Fig 69 A pick-up truck is a handy form of transport in the shellfish industry, but is an extra expense

Loss by theft Gone are the days when the creel fisherman could leave his gear on the harbour side, safe in the knowledge that no matter how long it remained, the rightful owner could claim it at any time. Such action today, other than in the more isolated communities, is an invitation to thieves. Creels, rope, drums of fuel and even bait can vanish nowadays with monotonous regularity.

With the gear required for today's fishery and its

124

high volume, secure premises are needed both for stowage and making and repairing. Local authorities seem bitten by the bug of making every quay and harbourside into tourist attractions. Buildings long used by fishermen for stores and workshops, if not converted into arty craft shops, are likely to get the bulldozer treatment for conversion into car parks or flower beds.

The regular outlays Where the fisherman has no suitable premises around his home for storage and working, somewhere suitable will need to be rented or purchased.

Weekly, there will be the bill for bait to settle; even if it is self caught, there will be some expenditure on fuel and considerable time involved. Salt or ice for preservation cannot be ignored or skimped. A few extra pounds in this direction makes a difference in the quality of the bait going into the creel and the volume of the catch taken out.

Gear is continually wearing out and sooner or later requires replacement. Some spare gear in hand is a sound insurance against a total 'wipeout' by bad weather during the height of the lobster season. Dahns and rope are lost and their replacement can only be provided by dipping into profits.

Hull materials, if they are not to suffer from rot and rust, need at least an annual paint job. Even modern GRP boats after the first few years need the protection of a layer of paint. The totally maintenance free boat, like the perfect shellfish trap, awaits invention.

Over the year, numerous small jobs of hull and machinery require attention and expenditure. The older the boat the more frequent and costly these are likely to be. Routine servicing does cut the cost and likelihood of unexpected breakdown during a period of good fishing.

10 Marketing, care of the catch

In most branches of commercial fishing, periods of bad
weather with fleets laid up in harbour will bring a
dramatic increase in quay-side prices. This is much less
true in the case of shellfish, due perhaps to the catch
being marketed and then held alive in storage for
considerable periods. At least this is true in the case of
lobsters, and the trap fisherman soon learns that
working in poor weather does not increase his earnings
pro rata to the volume of the catch.

Annually, to all intents and purposes, the lobster
grounds seem devoid of any prospective catch as the
fish completes the ecdysis phase. When this is over and
they begin to emerge, the fish is at its weakest,
hungriest and most vulnerable when removed from the
sea.

Care needed in Clad in a new shell which is to all appearances hard
handling the catch and solid, the lobster at this time needs careful
handling if losses in the catch are not to result.
Moulted lobsters are immediately recognized by the
shiny clean barnacle-free appearance of their shells.
Pick them up any other way than gently and the shell
can be felt to give and bend inwards in the hand.

Essential to avoid dehydration When in this delicate state dehydration must be avoided at all costs. The drying out of the shell by sun or wind will cause a high rate of mortality in the catch. As there is no cure, prevention is essential. First the catch should if possible be stored in a shaded part of the deck – no easy matter on small boats during summer – then as the fleet is hauled and more lobsters are taken, they are placed in catch boxes with layers of well soaked hessian sacking over them. Wet sacking has a two-fold effect of both stopping drying out of the shell and at the same time preventing the catch embarking on a limb-tearing orgy of mutual destruction.

When placed in the confines of a fish box, lobsters become extremely aggressive towards their fellows. Two placed in a single container will back off and face each other up in preparation for battle. With little room to manoeuvre or retreat it will mean at the least a mangled or cast limb, at worst a dead lobster. Without doubt the best aid in lobster catch preservation is plenty of sea-water soaked hessian sacking.

Storage water must be aerated What should not be done with lobsters is to try to store them in any tank or barrel filled with water which is not being circulated and aerated. Lobsters will survive for up to twenty-four hours or even longer out of water as long as they are kept cool and moist; or they can be stored in circulating water for indefinite periods. A confined water supply with the available oxygen being used up, especially in warm weather, rapidly proves fatal.

Wet sacking is only a temporary measure to keep order among the catch while the crew is otherwise engaged. When the fleet has been reshot the catch should be gauged for size and those retained have their claws immobilized by thick rubber bands around the claws.

Immobilizing lobsters Originally, immobilizing was done by severing the claw

127

muscle or inserting a peg. With a high value catch such as lobsters this is an invitation to premature death among the catch – no one wants to buy a dead lobster. For this reason the rubber band method has been almost universally adopted, having the merit of being easily applied, effective and harmless to the catch. Care must still be exercised when fitting bands, always lifting the fish by the body, never the claws which are likely to be shed when grasped.

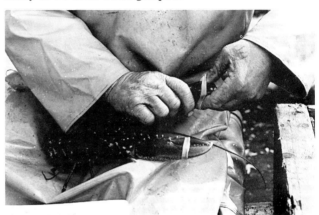

Fig 70 Securing lobster claws with heavy rubber bands

Storage of catch With the catch from one fleet safely immobilized, they should be stored on board in a lidded box, once again under a layer of wet sacking.

Crabs, being of a lower value than lobsters, never receive the special attention given to lobsters. The same basic rules apply of keeping them moist and cool and where possible away from the drying effects of sun and wind. On many boats the crab catch will simply be stored on deck.

When crabs are stored on deck in 'dry' conditions, deep boxes, barrels and tea chests have been traditionally used for this purpose; today there are now available plastic boxes made specifically for crab holding. Whether deep or shallow boxes are used, crabs should, rather than being thrown from several feet away, be placed in by hand, packing them down to minimize movement and therefore mutual damage.

Even for normal delivery to local processing outlets the crab should be subject to better handling than is sometimes given. Throwing crabs into boxes is bad practice as it causes damage to shell, claws, limbs and unseen internal injuries. All of these injuries can result in the premature death of the catch especially if they are not being collected for immediate processing locally.

Vivier tanks Larger boats with vivier tanks will store their catch live in circulating sea water after first immobilizing them by severing or 'nicking' the muscle used to close the claw. One of the main advantages offered by the vivier vessel is that catches can be landed live to ports offering the best markets. In Britain this will apply mainly to boats working in the English Channel, where continental harbours are within convenient steaming distance.

Exports to the continent With the increasing trend towards live export to continental markets from landing ports as far north as the Hebrides, even more care is required both in the handling and selection of crabs for this outlet. Selection is most important with only top quality hard crabs which feel heavy for their size, *ie* likely to produce a good meat yield picked for export.

Specimens which have damaged to carapace or apron are unlikely to survive the journey abroad and should be sent to local processing outlets or dumped. Most buyers for the live export market will have a specification to be met, generally requiring that both claws and most walking legs are intact.

Damaged legs can be 'repaired' by assisting the animal to shed them along the casting point where the legs join the body, for as in natural shedding during autotomy, the joint will seal itself.

Preparing crabs for transit To prevent crabs damaging each other during transit/and storage, as is the case with crabs going

129

direct into on board tanks, 'nicking' is essential. Using a triangular bar to hold the claw open, the muscle is cut with a special curve bladed knife. This of course leads to bleeding, but as crab claws cannot be secured with normal bands as in the case with lobsters, at the moment there is no alternative; but this is currently subject to further research including 'body banding' of the whole crab.

Damage of any kind to the crab, whether accidental or through knicking, will lead to the loss of blood and body fluids. In the stressful conditions of close storage, water will soon become contaminated leading to further deaths among the cargo.

Live transport of crabs is undertaken by specially adapted tanker lorries known as vivier trucks. In these the water is circulated, aerated, and in the most modern arrangements the temperature is controlled also. Even with these precautions, if the crab is not in first class condition at the start of the journey, it is unlikely to be alive at its destination.

Close co-operation necessary between catcher and buyer

Getting involved in live crab export is outwith the resources of individual fishermen. It requires much co-operation between catcher and buyer in the selection and treatment of crabs, and without prior planning and consultation successful exporting would be impossible to achieve.

Spanish and French buyers are interested only in crabs which arrive into their holding tanks 'in live', and this means lively healthy, condition. For consignments to arrive either dead or moribund will soon see that source black listed as unreliable.

Advisory publications

Fishermen and merchants who are considering entering upon live export should avail themselves of two publications from the Sea Fish Industry Authority's Industrial Development Unit. These are Technical Reports numbers 280 and 294 which deal with handling, transport and cause of mortalities among live crabs sent for export. Obtainable from Sea Fisheries House, 10 Young Street, Edinburgh EH2 4JQ or Industrial Development Unit, St Andrew's Dock, Hull HU3 4QE, where of course, further up-to-date advice is also available.

Prices Every year, every season, shell fishermen are given a multitude of reasons to reduce the price of their catch often when landings are not particularly good on their home grounds. High pound against dollar, low pound against dollar, the pound high or low against European currencies. Or it may be Canadian lobsters, north, south, east and west coast lobsters on the market – these are all reasons given at one time or another. When the price per pound of lobsters comes down, it is inevitably in multiples of ten pence per pound weight. When it rises it only ever seems to be one or two pence at a time.

Keep cages Where sheltered coves or inlets are available, it is possible for lobsters to be stored in keep cages until enough are collected for market. On a smaller scale, some lobsters can be retained in keeping creels, made in the same way as a catcher but without eyes or entrances. These creels can be bent to the main rope at the end of a fleet the same as an ordinary catching creel. Steel creels are preferred for keep creels, but if they are of wooden construction they must be kept well soaked and weighted to make sure they always settle base down. Small-mesh netting should always be used both for keeping predators out and preventing the escape of the occupants should one of the meshes get rubbed through.

Necessary precautions
for sea-water tanks The majority of the lobsters caught in the UK have their first point of sale to merchants in coastal locations. In these premises there must be a means of keeping lobsters alive for a period of from several days up to two or three weeks. This live storage takes the

131

shape of sea-water ponds or tanks in secure premises.

Usual practice is for the storage area to be divided into several different compartments. This serves a twofold purpose as the catch can be graded by size into different compartments, more convenient when packing lobsters to forward to their next point of sale; also it reduces the risk of mass hysterical deaths of the occupants which has been known to occur for unexplained reasons.

There are other more explicit ways in which the entire stock of holding ponds can meet an expensive and untimely death. Usually this can be attributed to the failure of some part of the system or machinery serving the installation. To keep the stock alive and healthy there must be a continuous supply of unpolluted, continually circulated, aerated and filtered, sea-water. This is mainly drawn directly from the sea, but inland storage is now possible with a correct mix of suitable chemicals which are now available.

Modern filtration systems are being developed which eliminate the need for frequent changes of water in lobster-holding ponds. These systems incorporate limestone filter beds and devices for removing harmful proteins and nitrates from the water.

Due to the amount of money which can be tied up in a stock of lobsters, it is without doubt a good plan to duplicate all pumps supplying the system. Even the use of a standby generator for the event of total power failure is a sound investment. Such is a fact of life – a system will seldom break down when those who are responsible are in attendance. An alarm system, either direct or by telephone link to the home of whoever is responsible for the system, can lead to an early rectification of any faults which occur.

All in all, it can be an expensive operation requiring frequent draining and refilling of the tanks with fresh sea-water. While the physical risks are negligible, the financial ones are not so, and many a shining star in the shellfish market has risen and sunk without trace. The investment needed for setting up a system to store lobsters, even from one boat at the height of the season, is beyond the financial means of most fishermen – the reason why the majority of marketing is through merchants rather than by direct sales.

This is where the fisherman is in a cleft stick. At some times of the year he will have insufficient lobsters to supply a local market. At other times a surplus, without storage facilities, will be an embarrassment when having to dispose of them through merchants.

Marketing outlets   From the coastal merchants the lobsters are marketed through wholesalers in major cities throughout the UK; London's Billingsgate being the principal outlet in this respect, where many top-class restaurants obtain their lobsters. Just as likely, the British-caught lobster may end up on a market in France or Belgium. Transport can be either by a vivier wagon, with continuous circulating sea-water, or, more usually being packed in wood shavings or wet newspaper, well iced down inside insulated boxes. From some of the more distant catching points on the Scottish islands, some times of the year the lobster are flown south as air freight.

Edible crabs have a wider market than the highly priced lobster flesh, as well as being considered by some people to be the better flavoured of the two species. Fortunate indeed are the fishermen who operate near large centres of population where a ready market for crabs is available. The same crab which will fetch only £3 to £4 per stone from a merchant can by the simple process of scrubbing and boiling be transformed into a product worth £1 or more per pound weight. For the cost of fuel and a little labour the fishermen can realize a profit from his catch which would normally have gone to others.

Transit containers   A new insulated container chilled by liquid nitrogen is available from GRP Cooltrucks Ltd of Leeds. This

133

easy clean unit measuring 1900 mm × 620 mm × 970 mm is suitable for loading with fish on trays at the quayside and can be delivered direct to supermarkets or other outlets. The SFIA has carried out tests on these units and initial trial runs appear to be favourable.

The 'Cooltrol' as the container is called, holds fourteen standard supermarket trays and has a temperature tolerance of $0 \cdot 3°$C.

The units are priced at between £1,000 and £1,200 with overnight delivery rates by British Rail 'Red Star'. The Goodall Partnership of Regent Street, London, have been responsible for the development of 'Cooltrol' and intend to offer a leasing and management scheme once the system gets under-way.

Certainly this 'Cooltrol' seems to offer a useful method for the small supplier to get high value shellfish to market without the need of a middle man and his cut of the profits. If British Rail rates are competitive, it is a venture well worth further investigation.

Processing machines   Until quite recently, one of the drawbacks of the crab processing industry was the labour intensity of the job and the difficulty in getting the maximum yield of meat by hand picking methods. In the mid-1970s several machines were developed both here and in America for the mechanical extraction of crab meat, mainly that which was difficult to remove by hand from the body cavities and legs. Two main principles were used in the machines; centrifugal and brine separation. Both methods double the yield of white meat, but unfortunately at the same time some of the flavour is lost.

All crab processing is carried out with boiled crabs, which is the first step in the production line for crab products.

Cooking procedure   Before cooking, crabs should be 'drowned' in fresh *ie* ordinary tap-water. Water temperatures of around

$100°F$ will kill the crabs in anything from half to one hour, at $50°F$ the time may be as long as four or five hours. When larger crabs are to be cooked they are best killed with a spike or pointed rod inserted into the body just above the mouth. Death is almost immediate and is considered by animal welfare organizations to be more humane than drowning.

Crab flesh contains high numbers of bacteria, all rendered harmless by proper cooking and post-cooking care. Whether killed by drowning or spiking the crabs should be immediately cooked in boiling water for from 20 to 30 minutes, depending on the size. Dropping a basket of cold crabs into boiling water will cause an immediate drop in boiler temperature; cooking times need to be calculated from the time the entire water volume of the cooking vessel is at boiling point.

On completion of the boiling time the crabs are removed from the boiler and hosed down with cold water. From boiling time it takes up to four hours for the meat within the carcass to stiffen and set to give a maximum yield. Any longer storage time before processing should be done in a refrigerator to prevent the meat becoming contaminated.

Processing   Most large scale crab processing is done on 'production lines', with different operatives working on claws, shell and body meat. Where the fisherman is concerned he will be working on a small scale from home, offering perhaps dressed crab in addition to those which are only boiled.

While large-scale crab processing is a separate subject, the steps are basically the same for a small operation. First the shell and body is separated: a sharp thump on a hard surface with the edge of the shell being of great assistance. When the parts are detached the 'orange' segment parts, the gills, are discarded as is the gall bladder. For the remainder, the brown meat in the shell can be easily scooped out with

135

a spoon. The white meat in the body, once the legs have been broken off, are easily picked out with the blunt end of a teaspoon. It is the claws, especially those of the cock crab, which contain the majority of the prized white meat. To get at this the legs are disjointed and, where accessible, the meat is scooped out with a teaspoon. That which cannot be reached in this way needs the claw shell to be broken open with a small hammer or nut crackers.

Half the white meat, which should have been kept apart, is mixed with the brown with salt, pepper, mustard and a dash of vinegar. Placed back in a scrubbed oiled shell and decorated with the remainder of the white meat and there you have it – dressed crab.

It is the already mentioned labour-intensive task of butchering and extracting crab meat which is one of the main reasons why crab-based products have not gained a larger share of the shellfish market. At the moment the market consists mainly of the sale of live or cooked whole crabs, dressed crab and factory-processed crab paste. Experimental work by private firms and government departments is ongoing in an effort to provide more crab-based products. Crab *pâte*, seafood platters and barbecued whole crab claws are some of the products under current promotion.

11 Other work suitable for trap fishing boats

In The Lobster Fisheries of England and Wales[8] by R C A Bannister Fisheries Laboratory, Lowestoft the author states that the impression is given by local management bodies that the fisherman's perception of the parlous state of lobster fishing is a stock problem, and that alone. However, there is strong evidence that price may have played a particularly important part in generating the difficulty. The trend in average first sale lobster prices since 1983 was downward, although occasionally they recovered. The rate of decrease has clearly slowed in recent years. However, when corrected for the decline in the purchasing value of the pound the trend is clearly a declining one throughout. In real terms therefore the lobster fisherman has been particularly badly hit by this single factor, which may have been a prime factor in the removal of men from the industry.

The fisherman's dilemma Boiled down to the basics this means that to survive, the trap fisherman is going to either (a) increase his catch rate, (b) cut out dealers and take their share of profits, (c) indulge in other forms of fishing, or (d) leave the industry.

Despite declining earnings most men working at the

crab and lobster fisheries do so because notwithstanding all the drawbacks, they enjoy the job. Direct marketing has been dealt with in another chapter and is an option which will become more attractive to fishermen if the present downward trend in prices continues. Increasing the catch rate means fishing more creels; already from many ports the fishing is exploited to the maximum. Finding a suitable 'berth' to shoot can be a problem at the height of the season.

This leaves option (c) as a means to generate extra revenue for the boat and crew. Alternative creel fisheries exist around the British Isles for shellfish other than crabs and lobsters. The longest established of these is that for Norway Lobsters (*Nephrops norvegicus*) mainly based around the northwest of Scotland and the inner isles among the deep water in the sheltered sea lochs. This fishing is carried out with creels, roughly half the size of the standard east-coast creel and covered with small mesh netting. These creels are mainly of plastic coated steel construction and are fished at five fathom intervals on the main rope rather than the ten as for crab and lobsters.

Recently other trap fisheries have developed, mainly from the Channel ports and southwest England. The first of these being for spider crabs, previously thought of as a 'rubbish' species, but now finding a ready market in Spain. So partial are the Spaniards to the spider crab that they consider it worth while transporting them alive in vivier trankers.

In a similar way the velvet or swimming crab (*Macropipus puber*) and the green shore crab (*Carcinus maenas*) have also found ready markets in Spain. Fishing for the last two species takes place with gear similar to that used for the Norway lobster fishery but with steel-based creels for the harder ground worked instead of the netting as is used in that fishery.

To indulge in any of these fisheries the boat already rigged for trap hauling would need no further additions

138

other than the purchase of the appropriate gear.

Where the above species cannot be caught or there is no market for them, some other activities need to be considered if extra income is to be generated.

Boats should be fully utilized

Larger boats which are suitably equipped can turn to some form of mobile gear fishing when trap fishing is slack. Away from the shallow waters of southern England this requires a boat of a minimum 35 ft overall and having five or six foot draft to tow efficiently.

Smaller boats can work at gill netting or long lining where grounds for these fisheries can be found. These methods are suitable to most boats engaged in trap fishing, only requiring the installation of appropriate hauling gear for nets, most creel haulers being adaptable for the hauling of lines.

Charter parties

Another money earner which is suitable for boats from twenty-five foot upwards is the running of charter parties for angling or visits to offshore bird sanctuaries and islands. Before indulging in chartering it will be necessary to take out an appropriate licence through the local authority to keep within the law. One stipulation which local authorities make is the provision of adequate safety and fire-fighting equipment and a comprehensive insurance cover for passengers. Also galling for a fisherman who has sailed from port for many years: he may also have to take a navigation test to prove that he is a 'competent' person.

Much of the angling charter work involves Saturday and Sunday working when the majority of angling clubs wish to go afloat. Away from areas of large population mid-week bookings are difficult to pick up other than during the holiday season. Fortunately, when the majority of the shore-based populace take their holidays, it coincides with the tailing off of the crab fishing and as yet the lobster has not re-emerged

139

from ecdysis. This makes chartering an attractive proposition, especially as, even in a typical British summer, weather conditions are the most suitable for the job.

Advantages of trap fishing Despite the apparent and less apparent drawbacks, trap fishing is still a means of getting a toehold in the fishing industry. It remains possible on quite a limited budget to get afloat with sufficient gear to earn a full-time living. A retired lobster fisherman has referred to lobstering as a life sentence, hauling, making and maintaining gear and procuring bait. A six- or seven-day week during the period of peak fishing which in much of Britain is from April to October.

Starting trap fishing is less likely to lead to major skipper-inspired disasters such as hanging up a trawl and doors on wreckage. The worst which can happen with poorly placed gear is a negative catch. As has already been mentioned, there are plenty of natural disasters awaiting the participants of the trap fishery and their gear without the need for any which are man-made.

For several years now crab and lobster prices have failed to keep up with the cost of living and, as discussed at the beginning of this chapter, fishermen's earnings have fallen. With the increased interest shown in the export of live crabs to Spain and other EEC countries, with new crab products in the pipeline, and an apparent decrease in the Canadian exports, the future of the crab and lobster fishing does appear a little brighter than at any time during the past decade.

Fig 71 Crab and lobster
fishing areas of Britain

141

Appendix 1

Key to map showing peak times for crab and lobster fishing

Area (A)   Northeastern District from Berwick upon Tweed to Redcar

Female crab	Late April to July
Male crab	June to August
Peak landings	May and June

Lobsters	Emerge from ecdysis mid August; best fishing from August to October

Area (B)   Humberside District from Redcar to Donna Nook

Female crab	Late August, September and October
Male crab	April, May and June
Peak landings	May and June

Lobsters	Emerge from ecdysis July; best fishing August

Area (C)   Eastern District from Donna Nook to Thames

Female crab	March, April
Male crab	August to September
Peak landings	April to June

Lobsters	Emerge from ecdysis July; best fishing July to August

Area (D)   Southeast District from Thames to Lyme Regis

Peak landings

Crabs   No separate figures available for male and female, but peak
                     landings in July to September period

Lobsters   Peak landings May to July

Area (E)   Southwestern District from Lyme Regis to Fowey and Bude to Chepstow

Female crabs   September and October
Male crabs   February to May
Peak landings   September and October

Lobsters   Emergence from ecdysis variable; best fishing April, May and
                     June.

Area (F)   Western District from Fowey to Bude

Female crabs   May to October
Male crabs   April to November
Peak landings   September

Lobsters   Emergence from ecdysis April and May; but not consistent as
                     on East Coast; best fishing June and July

Area (G)   Wales District from Chepstow to Connah's Quay

Female crabs   September and October
Male crabs   April and May
Peak landings   August to October

Lobsters   Emergence from ecdysis inconsistent except Cardigan Bay
                     July; best fishing July

Area (H)   North Western District from Connah's Quay to Scottish Border

Female crabs   Late April to early August
Male crabs   Late April to early August
Peak landings   May

Lobsters   Emerge from ecdysis late July and August; best fishing
                     August and September

Area (I)   Ayr District from Solway to Clyde

Female crabs	Clyde; October to November	Solway; April to June
Male crabs	Clyde; June to July	Solway; May to June
Peak landings	Clyde; July to August	Solway; May to June
Lobsters	Emerge from ecdysis	
	Clyde; June to July	Solway; July to August
Best fishing	Clyde: July to September	Solway: August to September and April to May

Area (J)   Campbeltown District from Tarbert to Crinan

Female crabs	October to December
Male crabs	All year, similar trend to females
Peak landings	October to December

Lobsters   Emerge from ecdysis June to August; best fishing June to August.

Area (K)   Oban District

Female crabs	September
Male crabs	August and September
Peak landings	August to October

Lobsters   Emerge from ecdysis June; best fishing August and September

Area (L)   Mallaig District

Female crabs	August and September
Male crabs	February to April
Peak landings	August and September

Lobsters   Emerge from ecdysis May; best fishing August and September.

144

Area (M)   Ullapool District from Gairloch to Achiltibuie

Female crabs	May to September
Male crabs	October to April
Peak landings	September to November
Lobsters	Emerge from ecdysis in July; best fishing August to November

Main effort in this District is concentrated upon Nephrops.

Area (N)   Kinlochbervie District Badcall Bay (Scourie) to Strathy Point

Female crabs	August to October
Male crabs	August to October
Best landings	August to October
Lobsters	Emerge from ecdysis July; best fishing October to November

Area (O)   Stornoway District Outer Isles

Female crabs	July to October
Male crabs	December to March
Peak landings	August to September
Lobsters	Emerge from ecdysis June to July; best fishing August to October

Area (P)   Wick District

Female crabs	July to August
Male crabs	May to June
Peak landings	July to August
Lobsters	Emerge from ecdysis August; best fishing September

Area (Q)   Orkney Islands

Female crabs	August
Male crabs	May to June
Peak landings	August
Lobsters	Emerge from ecdysis August; best fishing September

Area (R)   Shetland District

Female crabs   March to April
Male crabs   September
Peak landings   July

Lobsters   Emerge from ecdysis June to July; best fishing September to
                 December

Area (S)   Buckie District from Spey Mouth to Cullen

Little direct crab fishing takes place in the immediate vicinity to Buckie itself,
main landings being from part-time fishermen during the pursuit of lobsters in
summer.

Lobsters   Emerge from ecdysis late July; best fishing August to October

Area (T)   Macduff District from Cullen to Troup Head

Female crabs   July to September
Male crabs   May to July
Peak landings   August to September

Lobsters   Emerge from ecdysis variable; best fishing July to September

Area (U)   Fraserburgh District from Troup Head to Rattray Head

Female crabs   May to August
Male crabs   Very few landed
Peak landings   May to August

Lobsters   Emerge from ecdysis July: best fishing August to September

Area (V)   Aberdeen District, Bervie Bay to Aberdeen

Female crabs   July to September
Male crabs   May to August
Peak landings   June to August

Lobsters   Emerge from ecdysis August; best fishing September to
                 November

146

Area (W)   Arbroath District from Inverbervie to Carnoustie

Female crabs   Mid-April to May
Male crabs   Mid-August to October
Peak landings   May

Lobsters   Emerge from ecdysis September; best fishing September to
                October
                Date variable dependant upon water temperature.

Area (X)   Pittenweem District, from Perth to Alloa

Female crabs   May and June
Male crabs   August and September
Peak landings   May and June

Lobsters   Emerge from ecdysis August; best fishing September to
                October

Area (Y)   Eyemouth District from Alloa to English Border

Female crabs   May to July
Male crabs   June to August
Peak landings   May to July

Lobsters   Emerge from ecdysis August; best fishing August to October

These maps and keys have been compiled from information supplied by Fishery
Officers throughout the country. While giving a general trend, no account has
been taken of any seasonal variations in weather and/or sea conditions which
can affect crab and lobster landings.

  In some areas such as SE Scotland there is a general trend for crab and
lobster seasons to become later in the calendar year. Ten years ago crab fishing
was at its height in May, today (1988) it is June before peak landings of crabs
take place.

  Lobsters show a similar trend, with the emergence from ecdysis now more
likely to be mid-August rather than the last week in July.

  Again, even within areas there may be parts of the coastline where catches of
shellfish are excellent at certain periods, while a few miles away landings are
poor. To take all these into account throughout Britain would require a book in
its own right, therefore, while the map has been compiled in good faith it is in
no way a bible to every bay and inlet where crab and lobster fishing takes place.

147

Appendix 2

Conversion table from pounds to kilograms taken to second decimal point

Pounds	Kilograms
1	0·45
2	0·90
3	1·36
4	1·81
5	2·26
6	2·72
7	3·17
8	3·62
9	4·08
10	4·53

Conversion tables of inches to millimetres taken to second decimal point

Inches	Millimetres
1	25·4
2	50·8
3	76·2
4	101·6
5	127·00
6	152·4
7	177·8
8	203·2
9	228·6
10	254·00

Conversion tables of fathoms to feet to metres taken to second decimal point

Fathoms	Feet	Metres
1	6	1·82
2	12	3·65
3	18	5·48
4	24	7·31
5	30	9.14
6	36	10·97
7	42	12·80
8	48	14·63
9	54	16·45
10	60	18·28

References

1. Anon, The lobster: its biology and fishery. Ministry of Agriculture, Fisheries and Food, Fisheries Laboratory, Burnham on Crouch. 21.4.77 L.10
2. Howard, The influence of topography and current on size composition of lobster populations. Ministry of Agriculture, Fisheries and Food, Fisheries Laboratory, Burnham on Crouch. CM 1977/K:31 Shellfish and Benthos Committee.
3. Howard, A E and Bennett, D B, 1979. The substrate preference and burrowing behaviour of the lobster, *Homarus gammarus* (L). J. Nat. Hist. Vol 13 pp 433–438.
4. Howard, A E and Nunny, R S, 1983. Effects of near-bed current speeds on the distribution and behaviour of the lobster, *Homarus gammarus* (L). J. Exp. mar. Biol. Ecol. 71; 27–42.
5. Edwards and Early, Catching, handling and processing crabs. Torry Advisory Note No 26 (revised).
6. Shelton, R G J, How lobsters get into creels. Scottish Fisheries Bulletin 46, 1981, Department of Agriculture and Fisheries for Scotland.
7. Shelton, R G J and Hall, W B, A comparison of the efficiency of the Scottish creel and the inkwell pot in the capture of crabs and lobsters. Marine Laboratory, Aberdeen. Fisheries Research 1 (1981/1982) 45–53.
8. Bannister, R C A, The lobster fisheries of England and Wales. Fisheries Laboratory, Lowestoft. Ministry of Agriculture, Fisheries and Food, Internal Report.

Glossary

Anchor	Either a proper anchor used to hold either end of fleet or can be a heavy piece of metal.
Automatic hauler	Creel hauler which hauls rope onboard unattended.
Autotomy	The ability of shellfish to cast a limb to facilitate escape. The limb regenerates.
Backrope	Main rope off fleet of creels or pots, to which short beckets are tied.
Bait	Fish or offal inserted in creel to encourage shellfish to enter.
Bait band	Doubled twine or rubber tube for holding bait in creel.
Bark	Tannin which is used as a preservative for wooden creels.
Becket	Short length of rope joining creel to backrope or messenger.
Berry hen	Female crab or lobster carrying eggs.
Bow or bough	Usually describes stem of boat but is also used to name curved part of framework of creel. When applied to traditional creel it is unclear as to whether this or bough is the correct term.
Bend	To tie.
Bull	Heavy transverse part of wooden creel base.
Bunch	Describes the state of a heap of static gear rolled into a heap by heavy weather.
Bridle	*See* becket and strop.

151

Capstan	Drum used for hauling creel rope.
Carapace	Part of lobster containing digestive and reproductive organs.
Casting	Fishermen's term for ecdysis process.
Creel	Traditional oblong shellfish trap with semi circular frame, locally known as crave, creeve and crib.
Carvel	Form of boat hull construction where planking abuts.
Clinker	Form of hull construction where planking overlaps.
DAFS	Department of Agriculture and Fisheries for Scotland. Scottish equivalent of MAFF.
Dahn	Marker used to indicate position of static gear. Comprising rope, buoy and flag.
Decca	Electronic navigation instrument receiving signals from shore stations which are displayed as positions in 'lanes'.
Dip	To pass one rope under the other when creels from two different boats have been shot across each other.
Door	Opening part of creel where catch is removed and bait inserted.
Ecdysis	The periodic shedding of the shell in the process of growing.
Echo sounder	Electronic machine sending and receiving sound signals displaying in graphic form the nature of the sea-bed.
Ejector	Metal part used to clear rope from vee wheel of automatic hauler. Also known as stripper.
Eye	Opening in a creel where shellfish can enter.
Faked	Gear shot in a zigzag fashion to drop as many traps as possible on small patch of productive ground.
Fast	Gear held by sea-bed obstruction.
Fastener	Sea-bed obstruction.
Fleet	Where more than one creel is worked on a common back rope or messenger.
Foul shoot	When creels are placed overboard in an improper order.
GRP	Glass reinforced plastic. A synthetic laminate popular in the construction of boats suitable for creeling.

Ground	Description of sea-bed, meaning a favourable type of bottom for the species being sought.
Hard	Sea-bed composed of rocks, stones or boulders.
Hydraulics	System basically comprising pump and motor capable of powering haulers and other deck machinery.
Jeannie	*See* roller fairlead.
Keep cage	A sunken box or large creel used to store shellfish in sheltered, unpolluted waters.
Lath	Cross piece of wooden creel.
Lead	Where boat is pulled along the line of static gear by hauler or winch without the need to steam ahead. Can also mean a true line from rope or wire being hauled to hauler or winch.
Mark	A specific part of the sea-bed where position is determined by lining up fixed transits on land.
Moulting	*See* ecdysis.
MAFF	Ministry of Agriculture Fisheries and Food.
Marine engine	A propulsion unit designed specifically for marine use.
Marinised engine	A propulsion unit based upon vehicle or industrial engines suitably converted for marine use.
Mollologger	*See* roller fairlead.
Main rope	Heavy rope to which creels or pots are tied at intervals.
Mobile gear	Fishing equipment which is towed by one or more boats.
Over-running	Method of working shellfish traps where only a few from a fleet are taken onboard at any one time.
Parlour creel	Creel with an extra compartment intended to make escape more difficult.
Pond	Enclosed area equipped with circulating sea water for the storage of live lobsters.
Pot	Type of shellfish trap used in the south of England, known as 'inkwell' pot due to shape. Once made from willow but today mostly steel and/or plastic.
Radar	An electronic machine which transmits and receives radio signals which are reflected by obstacles, on a screen. (Coastline etc.)

Roller fairlead	Fitting at gunwhale having vertical and horizontal rollers to guide a rope to a hauler.
SFIA	Sea Fish Industry Authority. 'Seafish' (in short) charged by the government to administer, among other functions, financial assistance to the fishing industry.
Side stick	Part of framework of traditional creel.
Slat	Cross piece of wooden creel base.
Shooting	Placing creels overboard after cleaning and baiting.
Shooting bar	Post or roller on gunwhale to control position of the rope running overboard when shooting.
Shooting iron	*See* Shooting bar.
Sonar	Electronic instrument transmitting and receiving sound waves underwater to determine nature of sea-bed or fish shoals.
Spinner	A plastic device to prevent beckets or strops being spun or unbraided during hauling and bad weather.
Strop	Short length of rope joining creel or pot to main rope, also known as strap.
Stripper	Metal part used to clear rope from vee wheels of automatic hauler.
Tier	Alternative name for fleet.
Unbend	Untie.
Vee wheel	Concave wheels which grip rope on an automatic hauler.
Vivier	Today usually means a tank on board vessel or lorry equipped with temperature controlled circulating sea water for the transport of live shellfish.
VHF	Very High Frequency. A multi channel two-way radio with limited range.
Welldeck	Where the deck is watertight but some distance below gunwhale. Most are equipped with freeing ports to allow water to drain away but, to some extent, prevents any coming inboard through them.
White toed crab	A crab which has recently cast its shell and has not recovered sufficiently for marketing purposes.

154

List of Further Reading

See list of books published by Fishing News Books Ltd

Sea Fish Industry Authority (Technical reports 280 and 294)

Live handling, transport and cause of mortalities of live crustaceans

Available from Seafish, 10 Young Street, Edinburgh EH2 4JQ

Scottish Fisheries Bulletin 46, 1981
  Can Lobster Habitat be improved? How Lobsters get into Creels

Torry Advisory Note No 26 (revised)

Catching, Handling and Processing Crabs

A comparison of some Methods used in Lobster and Crab Fishing

Comparison of the Efficiency of the Scottish Creel and the Inkwell Pot in the Capture of Crabs and Lobsters

Available from Torry Research Station, PO Box 31, 135 Abbey Road, Aberdeen AB9 8DG

The Lobster: its Biology and Fishery L.10

The Influence of Topography and Current on Size Composition of Lobster Populations.

Lobster 'Seeding' A Promising Approach to the Problem of Increasing Natural Stocks.

Available from MAFF, Fisheries Laboratory, Burnham-on-Crouch CMO 8HA

155

Effects of Near Sea-bed Current Speeds on the
   Distribution and Behaviour of the Lobster, *Homarus
   Gammarus* (L.)
Fisheries Notice Number 44, Crab Migrations in the
   English Channel, 1968–75
Laboratory Leaflet (New Series) No 30 Norfolk Crab
   Investigations 1969–73
Available from Fisheries Laboratory, Lowestoft, Suffolk.

Index

Books published by
Fishing News Books Ltd

Free catalogue available on request

Advances in fish science and technology
Aquaculture practices in Taiwan
Aquaculture training manual
Aquatic weed control
Atlantic salmon: its future
Better angling with simple science
British freshwater fishes
Business management in fisheries and aquaculture
Cage aquaculture
Calculations for fishing gear designs
Carp farming
Commercial fishing methods
Control of fish quality
Crab and lobster fishing
The crayfish
Culture of bivalve molluscs
Design of small fishing vessels
Developments in fisheries research in Scotland
Echo sounding and sonar for fishing
The edible crab and its fishery in British waters
Eel culture
Engineering, economics and fisheries management
European inland water fish: a multilingual catalogue
FAO catalogue of fishing gear designs
FAO catalogue of small scale fishing gear
Fibre ropes for fishing gear
Fish and shellfish farming in coastal waters
Fish catching methods of the world
Fisheries oceanography and ecology
Fisheries of Australia
Fisheries sonar
Fishermen's handbook
Fishery development experiences
Fishing and stock fluctuations
Fishing boats and their equipment
Fishing boats of the world 1
Fishing boats of the world 2
Fishing boats of the world 3
The fishing cadet's handbook
Fishing ports and markets
Fishing with light
Freezing and irradiation of fish

Freshwater fisheries management
Glossary of UK fishing gear terms
Handbook of trout and salmon diseases
A history of marine fish culture in Europe and North America
How to make and set nets
Introduction to fishery by-products
A living from lobsters
The lemon sole
The mackerel
Making and managing a trout lake
Managerial effectiveness in fisheries and aquaculture
Marine fisheries ecosystem
Marine pollution and sea life
Marketing in fisheries and aquaculture
Mending of fishing nets
Modern deep sea trawling gear
More Scottish fishing craft and their work
Multilingual dictionary of fish and fish products
Navigation primer for fishermen
Net work exercises
Netting materials for fishing gear
Ocean forum
Pair trawling and pair seining
Pelagic and semi-pelagic trawling gear
Penaeid shrimps — their biology and management
Planning of aquaculture development
Refrigeration on fishing vessels
Salmon and trout farming in Norway
Salmon farming handbook
Scallop and queen fisheries in the British Isles
Scallops and the diver-fisherman
Seine fishing
Squid jigging from small boats
Stability and trim of fishing vessels
Study of the sea
Textbook of fish culture
Training fishermen at sea
Trends in fish utilization
Trout farming handbook
Trout farming manual
Tuna fishing with pole and line